PRAISE FOR *OPEN CIRCUITS*

"This book made me fall in love with electronics all over again."

—HAJE JAN KAMPS, TECHCRUNCH

"A celebration of the electronic aesthetic . . . blur[s] the line between engineering and art."

—ANDREW "BUNNIE" HUANG, AUTHOR OF
THE HARDWARE HACKER AND *HACKING THE XBOX*

"Its stunning cross-section photography unlocks a hidden world full of elegance, subtle complexity, and wonder. . . . *Open Circuits* has something for everyone to appreciate."

—LEE GOLDBERG, ELECTRONIC DESIGN

"An eye-catching and educational coffee table tome."

—GARETH HALFACREE, HACKSTER.IO

"Each page is both a dive into technological history and an ode to the evolution and aesthetics of electronics themselves."

—GRACE EBERT, COLOSSAL

"Anyone interested in electronics and/or macrophotography will enjoy this book from both an aesthetic and informational standpoint. . . . It's truly a technological and photographic masterpiece."

—JEREMY COOK, EMBEDDED COMPUTING DESIGN

Open Circuits

The Inner Beauty of Electronic Components

Eric Schlaepfer &
Windell H. Oskay

no starch
press®

San Francisco

Printed in China

Third printing

27 26 25 24 23   3 4 5 6 7

ISBN-13: 978-1-7185-0234-5 (print)
ISBN-13: 978-1-7185-0235-2 (ebook)

Publisher: William Pollock
Managing Editor: Jill Franklin
Production Manager: Rachel Monaghan
Production Editors: Rachel Monaghan and Hilary Mansfield
Developmental Editor: Nathan Heidelberger
Cover Design: Monica Kamsvaag and Susan Brown
Interior Design and Composition: Maureen Forys, Happenstance Type-O-Rama
Technical Reviewer: Ken Shirriff
Proofreader: Scout Festa

For information on distribution, bulk sales, corporate sales, or translations, please contact No Starch Press® directly at info@nostarch.com or:

No Starch Press, Inc.
245 8th Street, San Francisco, CA 94103
phone: 1.415.863.9900
www.nostarch.com

Library of Congress Control Number: 2022015611

About the Authors

Eric Schlaepfer runs the popular engineering Twitter account *@TubeTimeUS*, where he posts cross-section photos, shares his retrocomputing and reverse engineering projects, investigates engineering accidents, and even features the occasional vacuum tube or two. Some of his better-known projects include the MOnSter 6502 (the world's largest 6502 microprocessor, made out of individual transistors), the Snark Barker (a retro recreation of the famous Sound Blaster sound card), and the Three Fives and XL741 transistor-scale replica chip kits (available from Evil Mad Scientist Laboratories). His diploma—a BS in electrical engineering from California Polytechnic State University, San Luis Obispo—was signed by Arnold Schwarzenegger.

Windell H. Oskay is the author of *The Annotated Build-It-Yourself Science Laboratory* and is the cofounder of Evil Mad Scientist Laboratories, where he designs robots for a living. He holds a BA in physics and mathematics from Lake Forest College and a PhD in physics from the University of Texas at Austin. He has been shooting photos since high school and likes cats, except when their hairs become visible on electronic components at high magnification.

About the Technical Reviewer

Ken Shirriff restores old computers and electronics and writes about computer history. In his blog (*righto.com*), he looks inside everything from chargers to integrated circuits. Ken was formerly a programmer at Google and holds a PhD in computer science from the University of California, Berkeley. He has received 20 patents and added seven characters to Unicode. He frequents Twitter as *@kenshirriff*.

Contents in Detail

Acknowledgments

Special thanks to the following who helped with the creation of this book:

Thanks to John McMaster for decapping some of the chips that we photographed. Ben Wojtowicz graciously provided his old Nexus phone for us to tear apart. Ken Sumrall let us pick through his extensive HP calculator collection, looking for particularly photogenic LED displays. Greg Schlaepfer loaned us the vintage guitar amplifier. Ken Shirriff, besides inspiring us with his detailed technical teardowns, also reviewed the book for technical accuracy. Jesse Vincent provided us with some keyboard keyswitches that ultimately didn't make it into the book, but that we are nonetheless genuinely grateful for. Thanks also to Brian Benchoff for a circuit board that provided a great context shot. Thanks to Philip Freidin for some fruitful discussions. And to Lenore Edman for acting as a springboard for ideas and as an occasional hand model, and for allowing Windell to go on sabbatical for this work.

Thanks to the staff at No Starch Press for making this book a reality.

Finally, thanks to all the folks on Twitter for their enthusiastic responses to the original cross-section photos that inspired this book.

Introduction

Form ever follows function.

—LOUIS SULLIVAN, AFTER VITRUVIUS

We cradle our seamless phones in our hands like cool, river-smoothed stones. They feel pleasing to touch. One kind of phone might seem better than another not because of its technological merits but because of how it looks and feels. This is by design. This *is* design. Industrial designers, engineers, and artists spend countless hours adjusting every curve, color, and texture. Good design appeals to our physical senses—and ultimately, to our sense of elegance.

Less obvious is that every constituent part of our devices—every ELECTRONIC COMPONENT—is also an object that was designed. Many components are themselves composite devices, made up of even smaller parts, each of which represents countless hours of design and engineering.

In this book, we'll take a close look at a number of interesting electronic components. As we look at each, we'll learn a little bit about three things: how it works, how it was manufactured, and how it is used. But what makes these components interesting in the first place doesn't always fit within those three categories. Often it's simply enough that we *get to look at them.*

Sometimes the most mundane components reveal surprising artistry and intricacy. A nondescript rock, cracked by the geologist's hammer, reveals a geode of mineral brilliance. A hammer is a particularly apt metaphor since this book is an *unabashedly destructive* tour of electronics. To reveal what's inside, our tools included saws, sandpaper, solvents, polishing wheels, end mills, and (yes) an occasional carpenter's hammer.

To an engineer, an electronic component consists of three parts: its INTERFACE, ACTIVE AREA, and PACKAGING. The interface connects the component electrically and mechanically into a circuit, as with connecting wires and mounting holes. The active area makes the component useful. For example, there are doped areas of silicon in a transistor that allow it to amplify signals. Packaging provides structural support, environmental protection, and the external shape of the component.

Viewing a component as a sum of these three parts provides a useful perspective for understanding its *technical* design. Often, the active area is utterly dwarfed by the interface and packaging. That's absolutely reasonable in many cases, like when you want to make a tiny light-emitting diode—the size of a grain of sand—big enough to handle with your fingers.

It's another thing altogether to contemplate the *aesthetic* design of components. While teams of designers and artists collaborate on the outward appearance of consumer electronics, the same cannot be said about the outward appearance of each interior component. A typical smartphone owner will *never* see what the parts inside that phone look like.

This book is *not* about accidental design. Every single wire, resistor, capacitor, and chip we'll study was intentionally designed to meet specific technical needs, in terms of precision, usability, and cost. This book *is* about accidental beauty: the emergent aesthetics of things you were never expected to see.

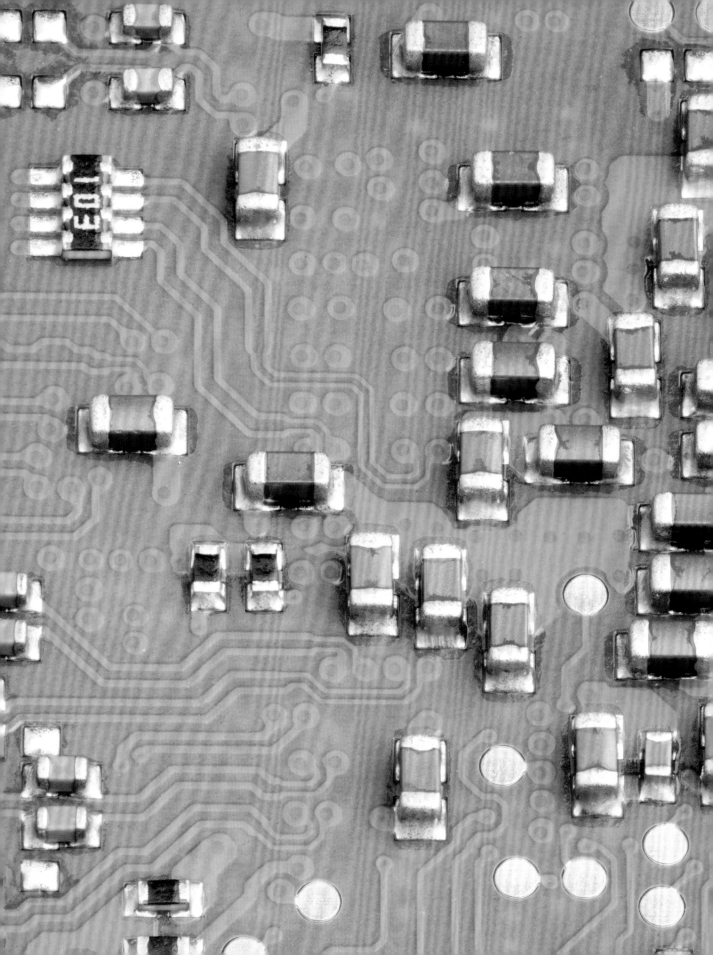

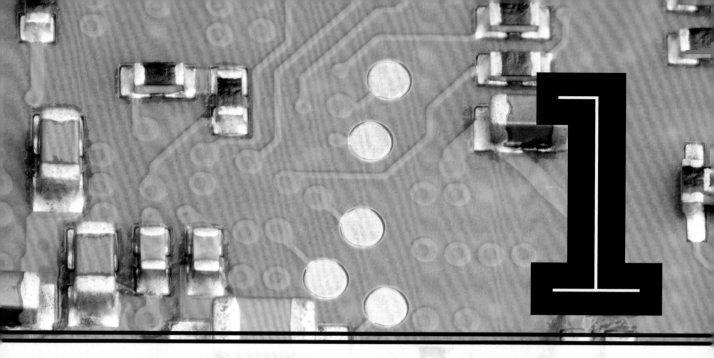

1

Passive Components

Resistors, capacitors, and inductors are basic building blocks found in essentially all electronic devices. All three are common examples of PASSIVE COMPONENTS, a broad category of components that don't add energy to a circuit. Instead, they dissipate, store, or transform energy in some way. These are some of the most varied and visually striking components, adorned with stripes, dots, glossy coatings, and cryptic labels. Let's take a look.

32 kHz Quartz Crystal

Deep within a quartz wristwatch lies a tiny tuning fork, cut from gleaming quartz crystal, that keeps the watch running on time. The tuning fork is plated with mirror-like electrodes on its surfaces and protected inside a tough metal tube.

A musician's tuning fork might be cut to ring at "A 440," the musical note A at 440 hertz (Hz), or 440 oscillations per second. The resonant frequency of this quartz tuning fork, however, is beyond the range of human hearing, precisely tuned to 32,768 Hz. (Keep dividing 32,768 Hz by 2 and you eventually get 1 Hz.)

Quartz is **PIEZOELECTRIC**: it flexes ever so slightly when a voltage is applied to it and also produces a voltage when flexed. The watch circuitry applies a tiny voltage to the electrodes, causing the quartz to flex and ring at its resonant frequency. As it does so, it produces an oscillating voltage. Every second, a digital circuit counts out 32,768 oscillations, then drives the second hand forward a single tick.

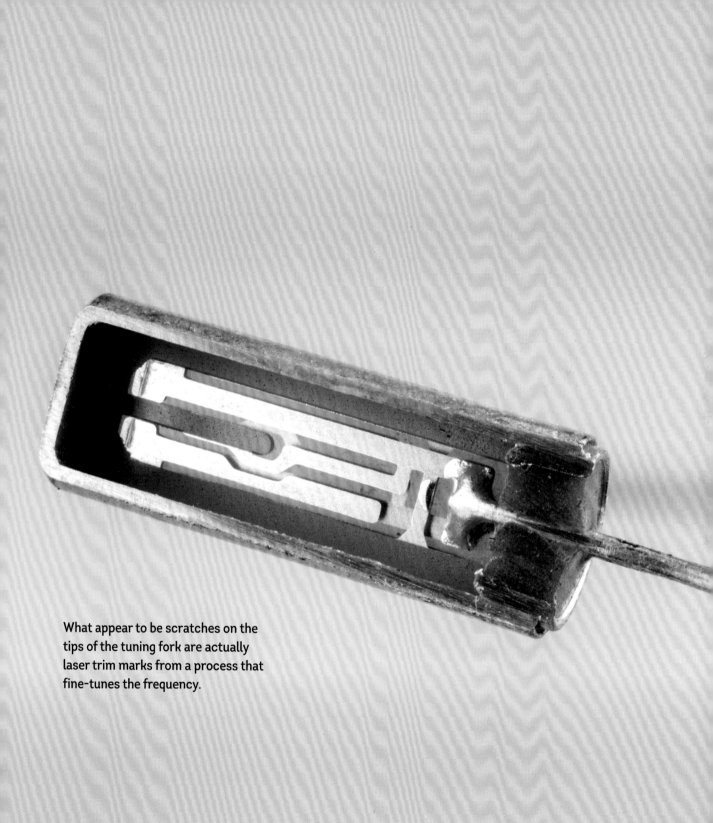

What appear to be scratches on the tips of the tuning fork are actually laser trim marks from a process that fine-tunes the frequency.

Carbon Film Resistor

RESISTORS are devices that restrict or limit the flow of electricity. They're used wherever a controlled amount of current is needed in a circuit. Everyday CARBON FILM RESISTORS like these are used in electronics like appliances and toys, where cost is more important than precision or size.

A carbon film resistor is made from a ceramic rod coated with a fine layer of carbon film that conducts electricity with some resistance. A helical groove is cut through the film, leaving a long narrow path of carbon that corkscrews from one end of the rod to the other. Metal caps are crimped onto the two ends, and wire leads are added. The resistor is then dipped in a protective coating and painted with color-coded stripes to indicate its resistance value.

Resistors of this shape are called AXIAL THROUGH-HOLE RESISTORS, meaning they have wire leads (intended to go *through holes* in a circuit board) arranged along the resistor's axis of symmetry.

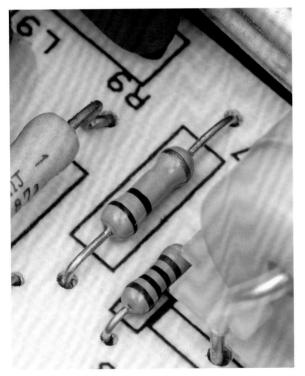

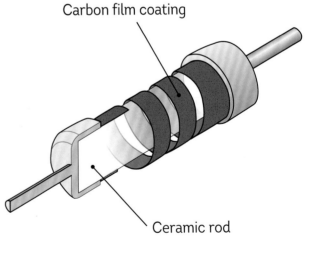

Carbon film coating

Ceramic rod

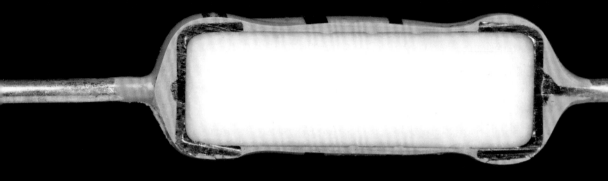

The carbon film is relatively thin. In cross section, the grooves are visible only as indentations in the ceramic rod.

The spiral groove can clearly be seen after the protective coating has been removed.

High-Stability Film Resistor

This **HIGH-STABILITY FILM RESISTOR**, about 4 mm in diameter, is made in much the same way as its inexpensive carbon film cousin, but with exacting precision. A ceramic rod is coated with a fine layer of resistive film (thin metal, metal oxide, or carbon) and then a perfectly uniform helical groove is machined into the film.

Instead of coating the resistor with an epoxy, it's hermetically sealed in a lustrous little glass envelope. This makes the resistor more robust, ideal for specialized cases such as precision reference instrumentation, where long-term stability of the resistor is critical. The glass envelope provides better isolation against moisture and other environmental changes than standard coatings like epoxy.

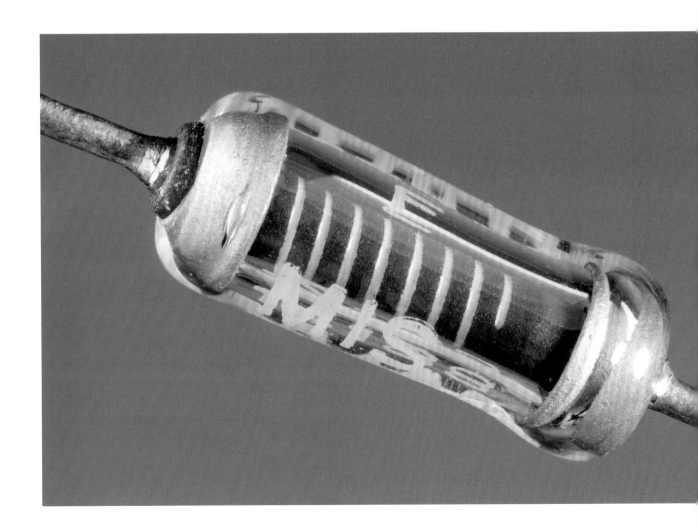

Wirewound Power Resistor

As current flows through a resistor, the resistor converts a certain amount of electrical energy into heat. Most general-purpose resistors have little ability to dissipate heat, as they cannot withstand elevated temperatures. This limits the amount of power they can handle.

POWER RESISTORS like this one are made without temperature-limiting materials like solder or epoxy, enabling them to handle more power. Some power supplies use them to limit the rush of current that occurs when you plug them in. The active element is a resistive metal wire wrapped around an insulating core. The resistive assembly is placed in a heat-tolerant ceramic shell and filled with cement grout.

The resistive wire is wrapped around a fiberglass core, but because this resistor is cut in half, all you see are the wire ends.

Thick-Film Resistor Array

Many circuits require multiple identical resistors. For example, a digital data bus might need a TERMINATION RESISTOR connected in series with each data line, or each I/O pin on a microcontroller might need a PULLDOWN RESISTOR between the pin and ground. A resistor array eliminates the need for multiple discrete resistors: it consists of several resistors fabricated as a single component.

Shown here are THICK-FILM ARRAYS, named for the fabrication technology, which uses silkscreened conductive and resistive films that are fired like pottery glazes onto a ceramic substrate.

After metal terminal pins are fitted and soldered, a laser burns away part of the resistive material to fine-tune each individual resistor to its correct specification. Finally, the array is dipped in an epoxy coating for protection.

This is a *single in-line resistor array*, or *SIL*, where all the terminals are arranged in a straight line. It has four independent resistors that aren't connected to each other.

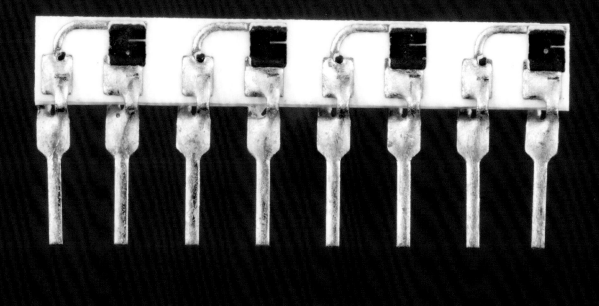

Straight cuts in the greenish resistive material mark the path of the trimming laser.

Surface-Mount Chip Resistor

The most common discrete resistors today are **THICK-FILM SURFACE-MOUNT RESISTORS**, also known as **CHIP RESISTORS** after their tidy rectangular packages, which lack wire leads. Billions of chip resistors are produced annually, and are found in every type of mass-produced consumer electronics.

These are **SURFACE-MOUNT** resistors, designed for soldering directly to the surface of a circuit board, as opposed to soldering wire leads that go through holes in the circuit board. They're constructed much like the resistors on thick-film arrays, right down to the laser trimming.

Several surface-mount chip resistors, with the epoxy coating peeled back to show the thick film element below

Thin-Film Resistor Array

THIN-FILM RESISTORS, such as the eight in this array, are precision devices manufactured by etching a pattern into an ultra-thin layer of **SPUTTERED** (vacuum deposited) metal oxide or **CERMET** (ceramic-metal composite). Thin-film arrays are used when a circuit requires precisely matched or calibrated resistors, such as for scientific or medical equipment.

Each serpentine track of resistive material has several areas that can be laser-trimmed to fine-tune the resistance value with increasing exactitude.

The solder ball terminals at the end of each resistor allow this array to be soldered directly to a circuit board.

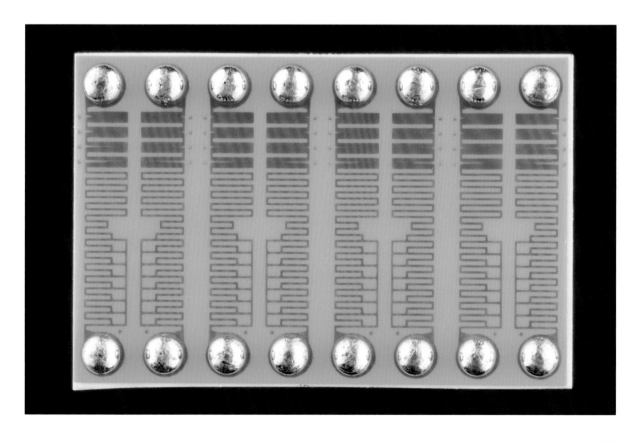

Wirewound Potentiometer

A **POTENTIOMETER**, or **POT**, is an adjustable resistor. Pots are used as front panel control knobs on everything from laboratory instrumentation to guitar amplifiers—anything where you turn a knob to adjust a setting.

This large pot is made of resistive wire wrapped around a ceramic form, an old design, essentially unchanged since 1925 and still in production today.

There are two terminals connected to either end of the resistive wire and a third terminal connected to a spring-loaded contact called a **WIPER**. The wiper touches the wire windings, making an electrical connection that can be moved around by rotating the shaft.

As the wiper moves away from or toward a terminal, the resistance between the wiper and that terminal increases or decreases because electrical current has to flow through a different amount of resistive wire. An amplifier circuit translates this changing resistance into a louder volume, or a hot plate interprets it as a temperature set point.

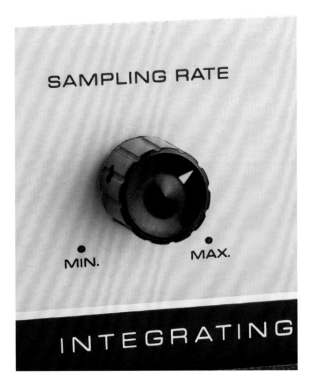

The wiper of a standard pot can be rotated about $^2/_3$ to $^3/_4$ of a turn, between the two fixed terminals.

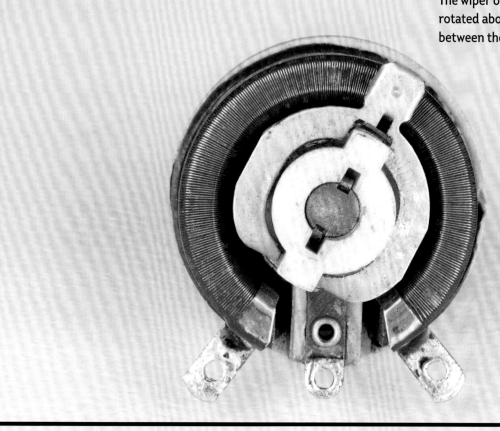

Most of the wrapped wire is covered with vitreous enamel, similar to a pottery glaze. Only the surface that contacts the wiper has exposed wire.

Trimmer Potentiometer

TRIMMER POTENTIOMETERS, often referred to by the trademark name TRIMPOT, aren't meant to be manipulated by the end user. Instead, they're designed for initial calibration and rare adjustment. You can find them in precision electronics that require fine-tuning at the factory or by service technicians. The typical service life of a trimmer is only a few hundred adjustments.

This colorful trimmer has a horseshoe-shaped section of resistive cermet film instead of a coil of wire. From the outside, you use a plastic adjustment tool or a screwdriver to turn the yellow plastic rotor. Inside, the rotor moves a flexible metal spring that acts as the wiper, connecting the center terminal to the resistive cermet film, changing the resistance between that center terminal and the two other terminals.

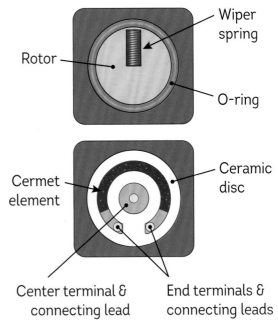

Rotor — Wiper spring

O-ring

Cermet element — Ceramic disc

Center terminal & connecting lead — End terminals & connecting leads

An orange O-ring underneath the rotor seals dust and debris out and provides friction to keep the rotor in place after adjustment.

15-Turn Trimmer Potentiometer

It takes 15 rotations of an adjustment screw to move a 15-turn trimmer potentiometer from one end of its resistive range to the other. Circuits that need to be adjusted with fine resolution control use this type of trimmer pot instead of the single-turn variety.

The resistive element in this trimmer is a strip of cermet silkscreened on a white ceramic substrate. Screen-printed metal links each end of the strip to the connecting wires. It's a flattened, linear version of the horseshoe-shaped resistive element in single-turn trimmers.

Turning the adjustment screw moves a plastic slider along a track. The wiper is a **SPRING FINGER**, a spring-loaded metal contact, attached to the slider. It makes contact between a metal strip and the selected point on the strip of resistive film.

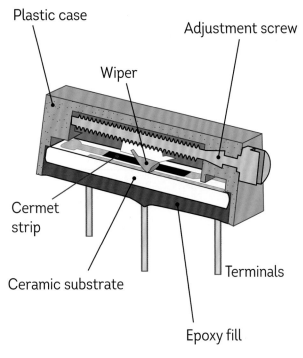

Plastic case

Adjustment screw

Wiper

Cermet strip

Ceramic substrate

Terminals

Epoxy fill

While it isn't obvious from the outside of the device, the adjustment screw is electrically insulated from all three pins of the component.

10-Turn Potentiometer

A 10-turn potentiometer is much like a wirewound pot, but its adjustment range is 10 full turns instead of less than one turn. This is a specialized device, occasionally found as an input knob on sensitive instruments where high adjustment resolution is required.

The wiper on a 10-turn pot keeps continuous contact with a helical track, moving up or down as the shaft is rotated. The track consists of resistive wire tightly wrapped around an insulated copper form. The wire ends connect to two of the terminals.

The connection between the wiper and the third terminal is through a vertical strip of brass that rotates with the shaft. As the wiper moves up and down, it maintains contact with the strip through a spring finger. Another spring finger keeps contact between the brass strip and the third terminal as the strip rotates.

The body of this pot was filled with clear resin to hold the contents in place while cutting it open.

Ceramic Disc Capacitor

CAPACITORS are fundamental electronic components that store energy in the form of static electricity. They're used in countless ways, including for bulk energy storage, to smooth out electronic signals, and as computer memory cells. The simplest capacitor consists of two parallel metal plates with a gap between them, but capacitors can take many forms so long as there are two conductive surfaces, called **ELECTRODES**, separated by an insulator.

A ceramic disc capacitor is a low-cost capacitor that is frequently found in appliances and toys. Its insulator is a ceramic disc, and its two parallel plates are extremely thin metal coatings that are evaporated or sputtered onto the disc's outer surfaces. Connecting wires are attached using solder, and the whole assembly is dipped into a porous coating material that dries hard and protects the capacitor from damage.

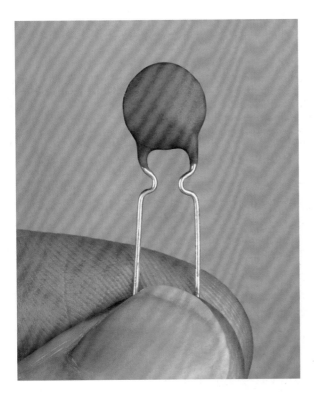

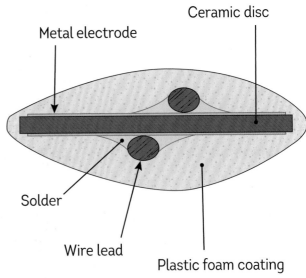

Metal electrode

Ceramic disc

Solder

Wire lead

Plastic foam coating

The metal layers on the surfaces of the ceramic disc are so thin that it can be hard to see them in cross section.

Glass Capacitor

A capacitor's amount of CAPACITANCE–
the electric charge it can store at a
given voltage–depends on the surface
area of the conductive plates, how close
together they are, and the type of insulator used between them. The insulator
is called a DIELECTRIC. While almost any
insulator–even air–can be used as a
dielectric, certain materials provide far
more capacitance than an air gap would.

This glass-packaged capacitor has multiple sets of aluminum foil plates interdigitated with each other. This layered
arrangement augments the available
surface area and increases the capacitance. Thin layers of glass, an excellent
insulator, function as the dielectric.

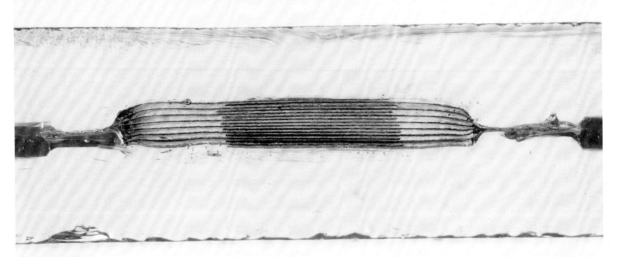

Eight foil layers on the left and eight
foil layers on the right are connected
to their respective terminals and are
precisely interleaved without touching.

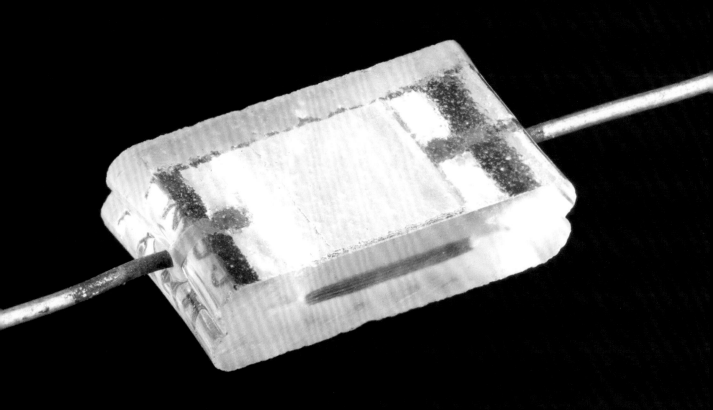

For robustness, the same type of glass
used between the foil layers is also used
as the outer packaging of the device,
about 5 mm thick.

Multilayer Ceramic Capacitor

MULTILAYER CERAMIC CAPACITORS (MLCCs) are the single most common discrete electronic component in production today; a smartphone may contain hundreds, most of which are used to ensure power supply stability at different points in the circuitry.

MLCCs are surface-mount CHIP CAPACITORS, consisting of interleaved layers of deposited metal between layers of specialized ceramic.

The one shown here in cross section is 1.5 mm long and has five interleaved metal layers, with two layers connected to one terminal and three to the other. Other MLCCs with different properties

may have thousands of layers in a device of the same size.

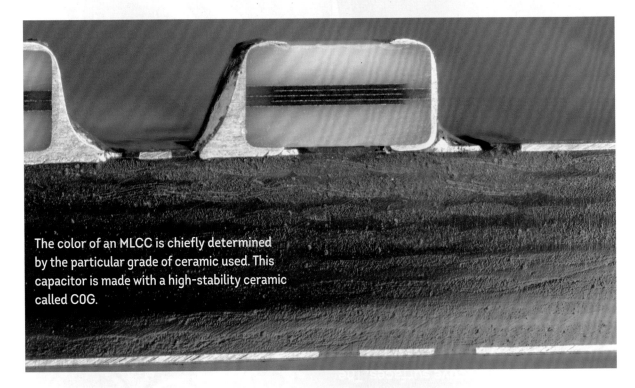

The color of an MLCC is chiefly determined by the particular grade of ceramic used. This capacitor is made with a high-stability ceramic called C0G.

Aluminum Electrolytic Capacitor

ALUMINUM ELECTROLYTIC CAPACITORS pack a large amount of capacitance into a small space, and are very common in power supplies. The outer metal can is filled with an **ELECTROLYTE**—an electrically conductive fluid. The fluid itself serves as one of the capacitor's conductive surfaces. The other is a long, thin, rolled-up strip of aluminum foil submerged in the fluid.

The aluminum foil is **ANODIZED**, producing an aluminum oxide on its surface that acts as the dielectric between the foil and the fluid. A second rolled-up aluminum foil strip, separated from the first by paper insulators, serves as a terminal, connecting the fluid to the wire leads.

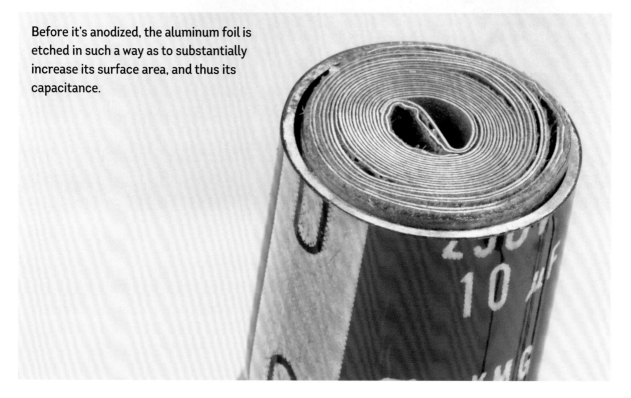

Before it's anodized, the aluminum foil is etched in such a way as to substantially increase its surface area, and thus its capacitance.

Film Capacitor

FILM CAPACITORS are frequently found in high-quality audio equipment, such as headphone amplifiers, record players, graphic equalizers, and radio tuners. Their key feature is that the dielectric material is a plastic film, such as polyester or polypropylene.

The metal electrodes of this film capacitor are vacuum-deposited on the surfaces of long strips of plastic film. After attaching leads, the films are rolled up and dipped into an epoxy that binds the assembly together. Then the completed assembly is dipped in a tough outer coating and marked with its value.

Other types of film capacitors are made by stacking flat layers of metalized plastic film, rather than rolling up layers of film.

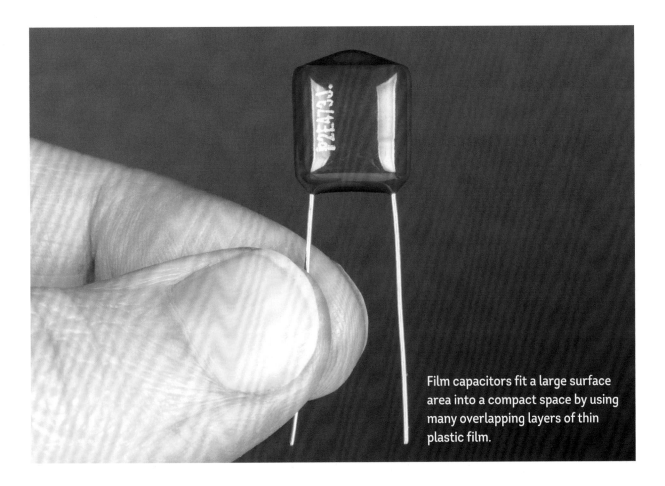

Film capacitors fit a large surface area into a compact space by using many overlapping layers of thin plastic film.

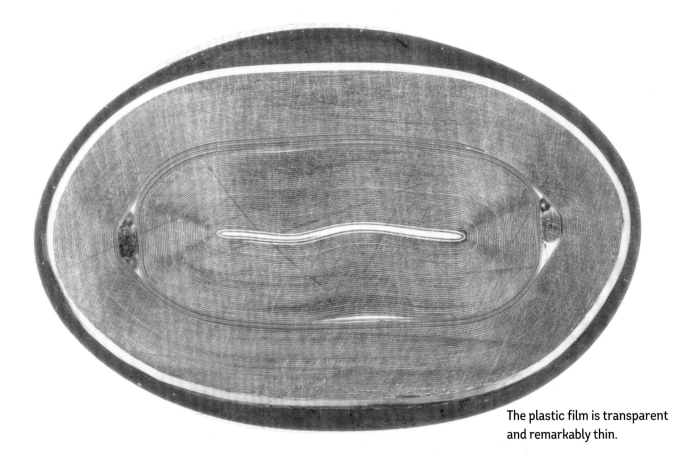

The plastic film is transparent and remarkably thin.

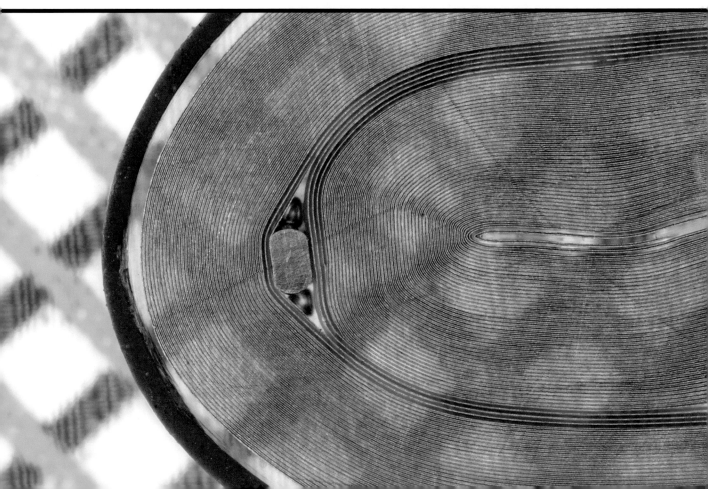

Dipped Tantalum Capacitor

At the core of this capacitor is a porous pellet of tantalum metal. The pellet is made from tantalum powder and sintered, or compressed at a high temperature, into a dense, sponge-like solid.

Just like a kitchen sponge, the resulting pellet has a high surface area per unit volume. The pellet is then anodized, creating an insulating oxide layer with an equally high surface area. This process packs a lot of capacitance into a compact device, using sponge-like geometry rather than the stacked or rolled layers that most other capacitors use.

The device's positive terminal, or **ANODE**, is connected directly to the tantalum metal. The negative terminal, or **CATHODE**, is formed by a thin layer of conductive manganese dioxide coating the pellet.

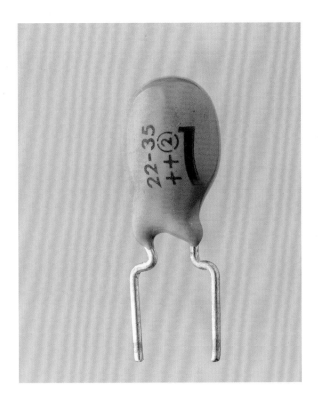

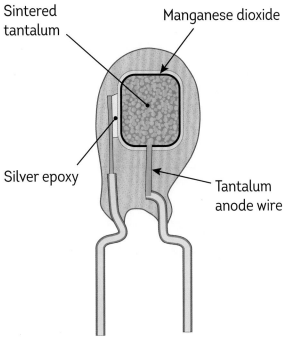

Sintered tantalum

Manganese dioxide

Silver epoxy

Tantalum anode wire

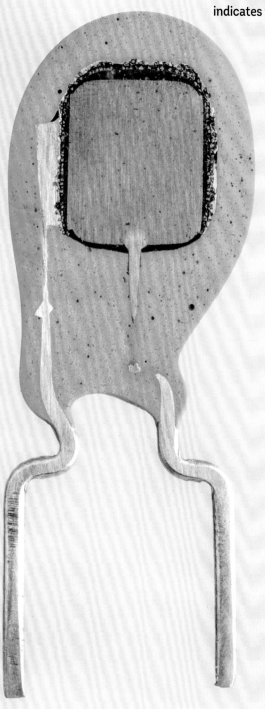

Connecting a tantalum capacitor backward causes chemical changes that damage the thin oxide layer. The label on the dipped plastic coating indicates the anode lead with "++."

Polymer Tantalum Chip Capacitor

POLYMER TANTALUM CHIP CAPACITORS are closely related to dipped tantalum capacitors. They're similarly based around an oxidized slug of tantalum metal with a high surface area. The slug is coated with a conductive polymer electrolyte, which flows into all its irregularities. Layers of carbon and silver paste connect the polymer to the cathode terminal.

The component is packaged in a molded epoxy case. It has tin-plated terminals for soldering to a circuit board. As a polarized device, it's labeled with both its value and a mark to indicate the anode.

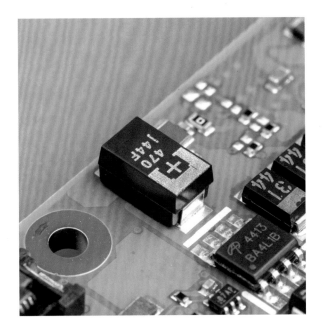

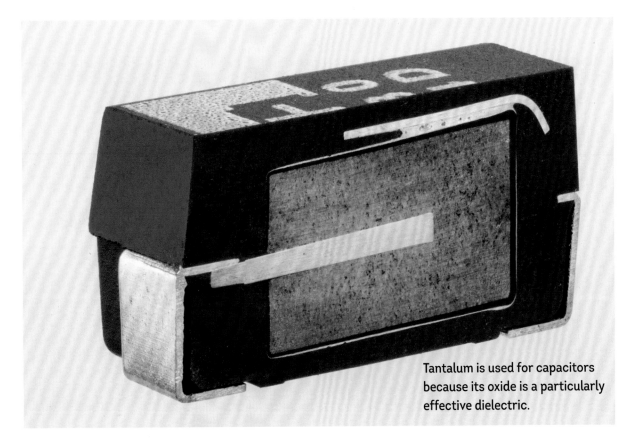

Tantalum is used for capacitors because its oxide is a particularly effective dielectric.

Polymer Aluminum Chip Capacitor

POLYMER ALUMINUM CHIP CAPACITORS are directly descended from standard electrolytic capacitors, despite how different they look both inside and out.

Instead of being rolled up, the etched and oxidized aluminum foils are laid flat and bonded together. And instead of a liquid electrolyte, this capacitor uses a conductive polymer as the cathode.

This newer style of capacitor is commonly found in smartphones, tablets, and laptops. Its popularity comes in part from its low profile, which allows it to fit where taller electrolytic caps cannot.

Layers of black conductive carbon paste and silver epoxy provide the electrical connection between the polymer-coated aluminum foils and the cathode terminal.

Axial Inductor

INDUCTORS are fundamental electronic components that store energy in the form of a magnetic field. They're used, for example, in some types of power supplies to convert between voltages by alternately storing and releasing energy. This energy-efficient design helps maximize the battery life of cellphones and other portable electronics.

Inductors typically consist of a coil of insulated wire wrapped around a core of magnetic material like iron or FERRITE, a ceramic filled with iron oxide. Current flowing around the core produces a magnetic field that acts as a sort of flywheel for current, smoothing out changes in the current as it flows through the inductor.

This axial inductor has a number of turns of varnished copper wire wrapped around a ferrite form and soldered to copper leads on its two ends. It has several layers of protection: a clear varnish over the windings, a light-green coating around the solder joints, and a striking green outer coating to protect the whole component and provide a surface for the colorful stripes that indicate its INDUCTANCE value.

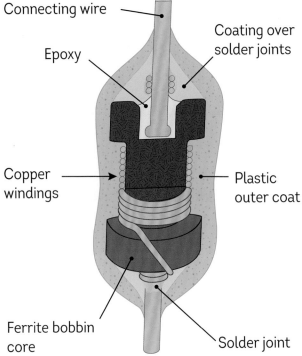

Connecting wire

Coating over solder joints

Epoxy

Copper windings

Plastic outer coat

Ferrite bobbin core

Solder joint

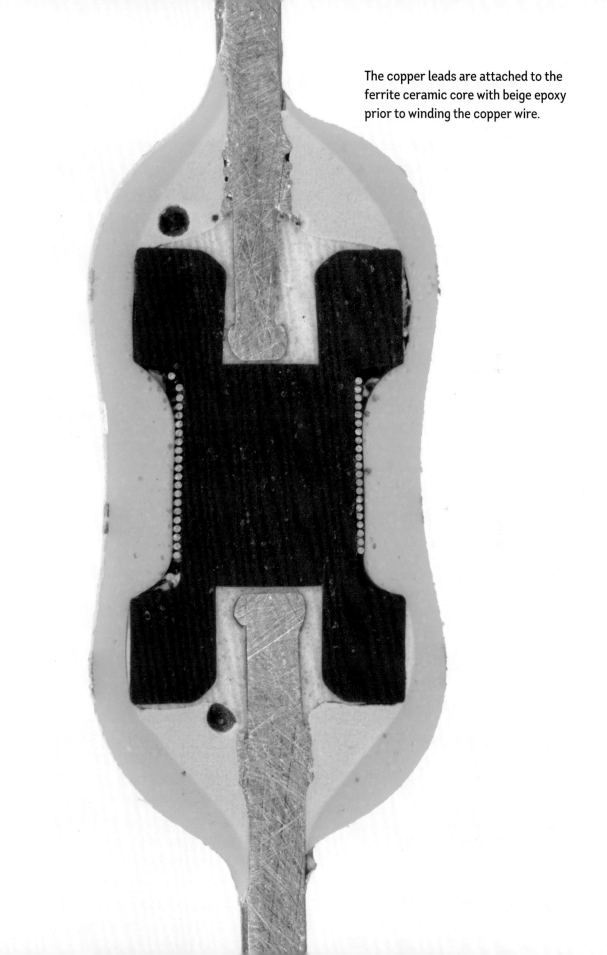

The copper leads are attached to the ferrite ceramic core with beige epoxy prior to winding the copper wire.

Surface-Mount Inductor

This surface-mount inductor, only 5 mm across, is designed to be compact, inexpensive, and easy for automated equipment to solder. You'd find it in cellphones, tablets, and laptops.

While axial inductors have wire leads that go through the circuit board, this inductor has terminals that sit directly atop the circuit for soldering.

The inductor has fine coils of varnished copper wire called MAGNET WIRE wound around a ferrite ceramic bobbin. The assembled core is placed inside another piece of ferrite to shield it from stray magnetic fields.

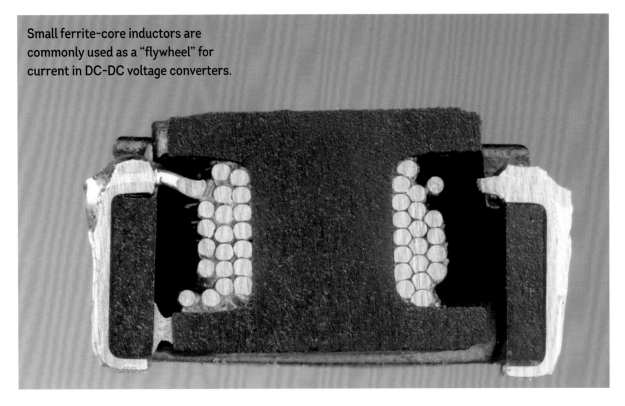

Small ferrite-core inductors are commonly used as a "flywheel" for current in DC-DC voltage converters.

Sintered Ferrite Inductor

This inductor, about 6 mm wide, has only two loops of copper wire. While we can't see it, the two ends of the looped wire are connected to the copper terminals on the left and right.

Unlike the other surface mount inductor, this inductor's copper windings appear suspended inside the solid ferrite as if by magic. It was manufactured in a sintering process: a fine ferrite powder was compressed into its final shape around the windings. Look closely and you'll see the copper windings have been pushed up against each other and deformed slightly as a result of this process.

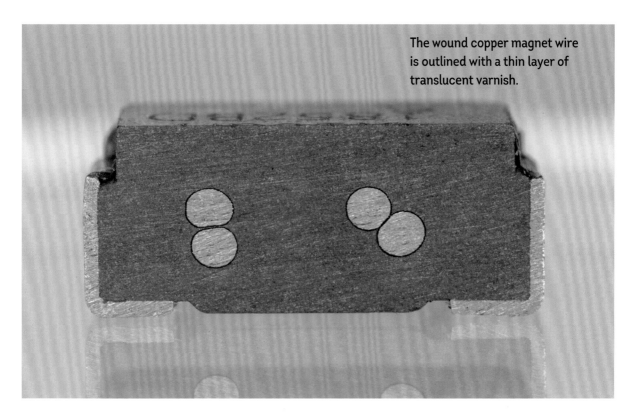

The wound copper magnet wire is outlined with a thin layer of translucent varnish.

Ferrite Bead

At first glance, this component may not look like an inductor at all. Where are the turns of wire? In fact, even a straight piece of wire with current flowing through it produces a magnetic field. The FERRITE BEAD surrounding this piece of wire just bumps up the inductance a little bit.

Ferrite beads can be used to stop stray radio waves from sneaking out of one electronic device and causing interference in another. They're also used to filter the power supply connections of sensitive chips, or to prevent electrically noisy chips from interfering with other chips on a circuit board.

This component is simple: just a bead of ferrite ceramic strung on a wire and glued in place.

Three-Terminal Filter Capacitor

This strange-looking component combines two inductors and a capacitor. A copper wire passes through two ferrite beads. Between the beads, one side of a ceramic capacitor is soldered to the wire. Another wire is soldered to the other side of the capacitor, forming the third terminal of the device.

Together, these parts act as a filter, preventing stray radio waves from wandering outside an electronic device and interfering with TV or Wi-Fi signals. Accordingly, you can find these devices on circuit boards next to connectors that go to the outside world.

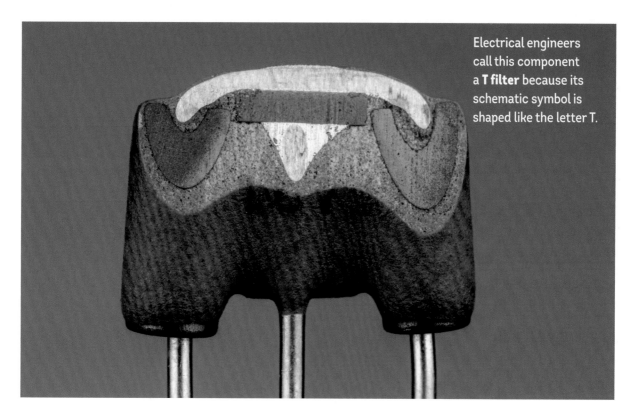

Electrical engineers call this component a **T filter** because its schematic symbol is shaped like the letter T.

Toroidal Transformer

A **TRANSFORMER** is an inductor wound with more than one coil of wire. The coils of these toroidal transformers are wound around donut-shaped ferrite cores.

Electrical current flowing through wire creates a magnetic field. Likewise, a changing magnetic field induces an electrical current in nearby wires. Thus when multiple coils are wound around a single core, changing the current in one wire changes the magnetic field, which creates a changing current in the other wire. This provides a method of **ELECTRICAL ISOLATION**: transmitting power or signals between wires without an electrically conductive path connecting them.

Having a different number of turns on the different windings can *transform* AC voltages from low to high or high to low. This kind of transformer is often used in power supplies for stepping up or stepping down voltages.

This transformer is configured as a choke: a special type of transformer designed to stop stray radio waves from leaking outside of a piece of electronics.

Power Supply Transformer

This transformer has multiple sets of windings and is used in a power supply to create multiple output AC voltages from a single AC input such as a wall outlet.

The small wires nearer the center are "high impedance" turns of magnet wire. These windings carry a higher voltage but a lower current. They're protected by several layers of tape, a copper foil electrostatic shield, and more tape.

The outer "low impedance" windings are made with thicker insulated wire and fewer turns. They handle a lower voltage but a higher current.

All of the windings are wrapped around a black plastic bobbin. Two pieces of ferrite ceramic are bonded together to form the magnetic core at the heart of the transformer.

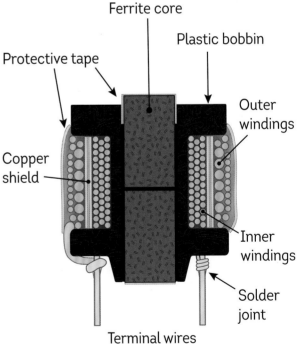

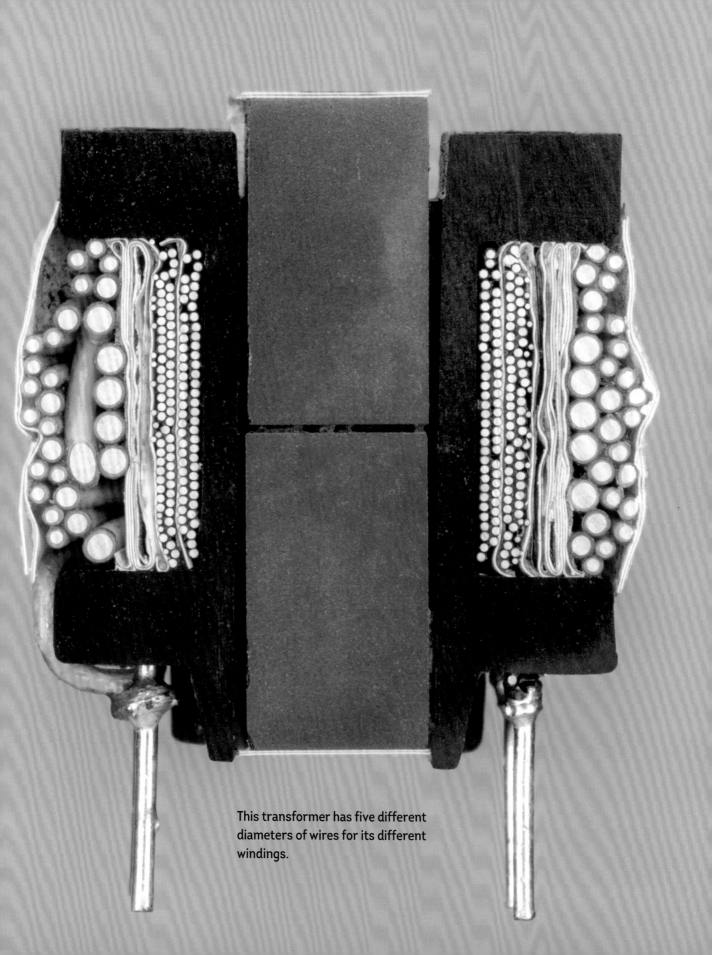

This transformer has five different diameters of wires for its different windings.

Low-Power Cartridge Fuses

FUSES are electrical components that break or "open" a circuit when more than a specified amount of electrical current passes through them, protecting other components from damage.

Here are a few glass cartridge fuses, each 0.25 inch in diameter. The two on the left are FAST-ACTING fuses rated for 10 and 15 amperes, respectively. They have round or flat metal wire between their ends. When the current exceeds the fuse's rating, the wire heats up enough to melt and quickly break the circuit.

The two fuses on the right are TIME-DELAY, or SLOW-BLOW, fuses, both rated for 0.25 A. Time-delay fuses resist spikes beyond their rating, requiring sustained current above that point to blow. One has thin wire wrapped around a fiberglass core that takes a while to heat up. The other has a resistor and spring. If the resistor overheats, it melts a dot of solder, releasing the spring and opening the circuit.

Fuses for very low current may have a fusible wire much thinner than even a human hair.

Cartridge fuses like these are found in equipment where the end user will replace the fuse. The glass case makes it easy to see when a fuse has blown.

Axial Lead Fuse

This component might look similar to a resistor, but it's actually a tiny fuse packaged with axial leads. Underneath the outer plastic coating is a ceramic tube containing the fuse wire. The wire is soldered to brass end caps pressed onto the copper lead wires.

This type of fuse is designed to be soldered to the circuit board, so it isn't meant to be replaced by the consumer. They're frequently used to provide additional protection to the circuitry if other protection circuitry fails.

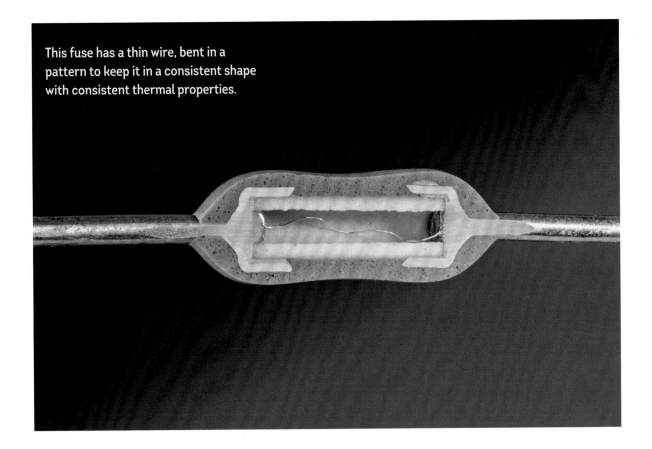

This fuse has a thin wire, bent in a pattern to keep it in a consistent shape with consistent thermal properties.

Liquid Power Fuse

At very high voltages, breaking a circuit gets tricky: long electrical arcs can easily form between pieces of metal as they separate, maintaining the flow of current. This huge, liquid-filled power fuse solves that problem.

Although the fuse has only a 15 A rating, it's designed to handle up to 23,000 volts. When the fuse is tripped, the long spring retracts below the surface of the liquid, pulling the broken ends of the fuse wire apart. The liquid insulates the end of the wire and quenches the electrical arc.

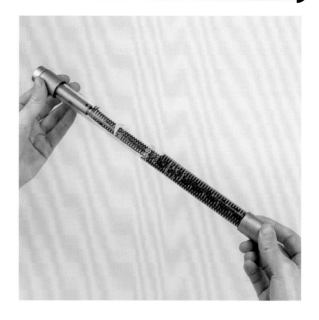

The liquid inside this vintage fuse is tetrachloroethylene, a chemical widely used as a dry-cleaning fluid.

Compact Power Fuse

Many handheld digital multimeters are protected from over-voltage and over-current conditions by compact power fuses. In fuses like these, an unexpected material surrounds the fusible element: grains of silica sand. The silica absorbs energy and quenches any electrical arc that might form when the fuse breaks, cutting off the current and ensuring that the circuit is fully disconnected.

Instead of a wire, these fuses contain a ribbon of metal, allowing them to handle higher currents. A dot of solder on the ribbon takes time to melt and thus acts as a simple time-delay element. A tough outer fiberglass tube protects the surrounding circuitry from the intense heat that can occur when the fuse blows.

Like the fluid in a liquid power fuse, the sand inside this fuse prevents arcing.

Thermal Fuse

A **THERMAL FUSE**, sometimes called a **THERMAL CUTOFF**, is like a regular fuse, except it opens an electrical circuit when it exceeds a certain temperature, rather than a certain level of current. Thermal fuses function as safety devices in electrical appliances that contain heating elements: coffee makers, hair dryers, rice cookers, and so on. They prevent a fire if some other part of the circuit fails.

The thermal fuse makes an electrical connection from one lead to the other via a spring wiper that contacts the edge of the metal case. The wiper is held in place by two springs braced against a wax pellet that melts at a specific temperature. When the wax melts, the springs expand into it, breaking the electrical connection irreversibly, even after the wax cools down and solidifies again.

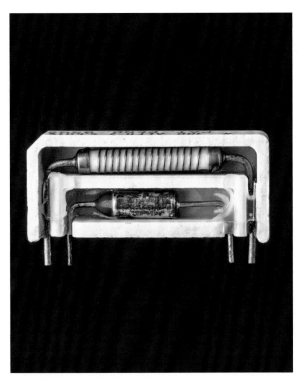

As a precaution, thermal fuses are sometimes packaged with wirewound power resistors, which are likely to be one of the hottest points in a circuit.

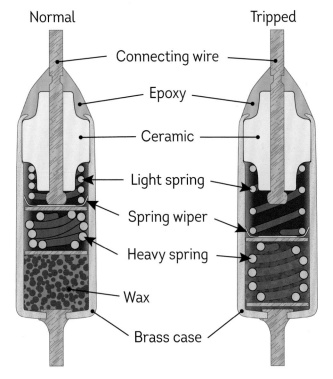

Normal

Tripped

Connecting wire

Epoxy

Ceramic

Light spring

Spring wiper

Heavy spring

Wax

Brass case

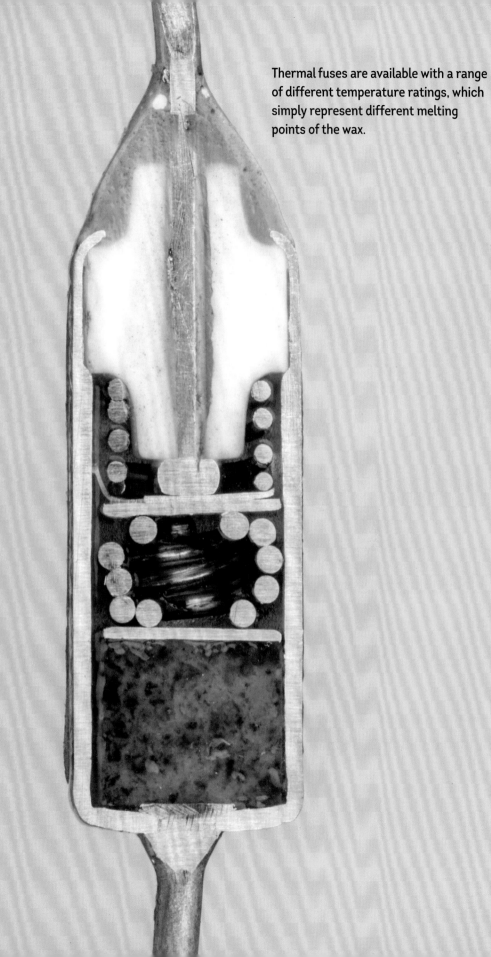

Thermal fuses are available with a range of different temperature ratings, which simply represent different melting points of the wax.

2

Semiconductors

Every area of modern life has been affected by the development of semiconductor devices. Light-emitting diodes (LEDs) now provide the light in our buildings and give life to animated billboards. Computer chips, camera sensors, and solar panels are also semiconductors. They work by exploiting the strange and wonderful electrical properties of ultra-pure crystalline materials like silicon, once they've been purposefully "poisoned" with minute amounts of impurities. Semiconductor components are often literal "black boxes" on circuit boards. Let's open some up and see what's inside.

1N4002 Diode

DIODES are components that allow current to flow only in one direction, much like a check valve in plumbing. They're commonly used in power supplies to convert alternating current to direct current.

The diode itself is a tiny "chip" of silicon, also called a DIE. The otherwise ultrapure silicon is modified to have separate regions: one where electrical current is carried by electrons, and another with HOLES, places where electrons are missing. The junction between the two regions, the active area of the device, can conduct current only in one direction.

In a 1N4002 diode, two large, tin-plated copper wires are bonded to the silicon chip with solder, and the result is encapsulated with a black epoxy plastic. The "ears" sticking out of the copper wires help retain them in the epoxy. The thin layers of solder connecting the silicon to the leads start out as thin disks that are melted during the assembly process.

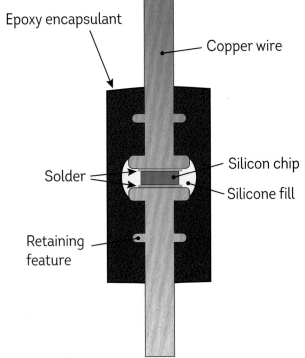

Epoxy encapsulant

Copper wire

Solder

Silicon chip

Silicone fill

Retaining feature

The white substance around the die is silicone rubber, much like regular household caulking. It protects the silicon chip during the assembly process.

Glass-Encapsulated Diodes

Low-power silicon diodes are often packaged in tubular glass envelopes, and they come in many varieties. In each, the actual silicon piece is tiny and is sandwiched between two contacts. Some diodes use a metal spring clip to make the connection between the silicon and one terminal. In others, both connecting wires are in direct contact with the die.

Typically, the outer glass envelope comes with one lead preinstalled.

Before being added to the component, the silicon die has a solder dot plated onto one side with an alternating-current electroplating bath; the diode itself ensures that current only passes in one direction, so the dot forms only on one side.

After the silicon chip is attached to the preinstalled lead by solder or conductive epoxy, the other lead can be attached.

A glass-encapsulated 1N740 diode on a circuit board

In this 1N914 diode, the tiny square die is positioned off center, possibly a manufacturing error.

The clear glass has been painted black on the outside of this 1N5236B Zener diode. An S-shaped spring makes contact with the die.

A tiny solder bump on the die of this 1N1100 diode connects it to the C-shaped spring contact.

Rectifier Bridge

The **RECTIFIER BRIDGE** might seem like a drab little hockey puck from the outside, but once the insulating plastic case is stripped away, an elegant circuit sculpture emerges. These components are commonly found in power supplies that plug into wall outlets. They consist of four silicon diodes, connected together in a special "bridge" arrangement that converts alternating current (AC) voltage into direct current (DC) voltage.

The four shiny gray silicon dies are sandwiched between the sets of wires. Two are facing up, and two are facing down, corresponding to the directions that current can flow within this little circuit.

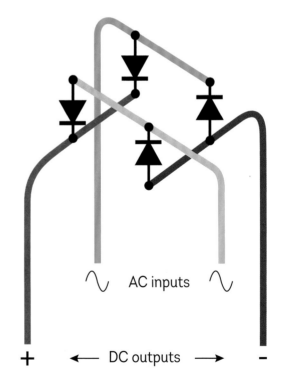

The wire leads are silver-plated copper. Their lower parts were tarnished by exposure to the air; the crisp, clean silver part was protected by the plastic package.

2N2222 Transistor

One of the key inventions of the 20th century, the **TRANSISTOR** is a semiconductor device that allows one electrical signal to control another. Transistors are commonly used to amplify signals or as logical switches.

The 2N2222 transistor shown here is a classic **BIPOLAR JUNCTION TRANSISTOR (BJT)** in a TO-18 metal "can" package. The active part is the shiny but tiny silicon die. By weight and volume, a device like this is *almost entirely packaging*.

A BJT has three terminals. Two, the base and emitter, are connected to the die through hair-thin aluminum **BOND WIRES** that reach from the ends of the insulated leads to the top of the die. The third connection is made through the bottom of the die to a third lead, the collector, that is electrically connected to the metal can. All three leads are held in place by a glass fill on the bottom of the device.

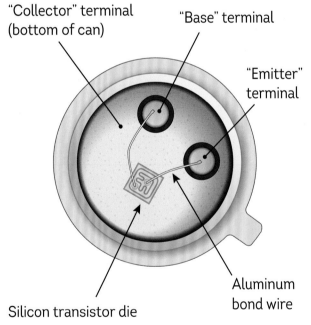

"Collector" terminal (bottom of can)

"Base" terminal

"Emitter" terminal

Silicon transistor die

Aluminum bond wire

2N3904 Transistor

This 2N3904 transistor is electrically similar to a 2N2222, but it looks very different because it's packaged in an inexpensive plastic case called a TO-92.

The active part of the transistor is a tiny gleaming die of silicon with three terminals, constructed much like that in the 2N2222. As in other BJTs, the base terminal controls the current that flows between the collector terminal and the emitter terminal, like a tiny electronic valve.

The black plastic, which constitutes the vast majority of the overall device, is molded from an epoxy filled with silica. All materials, including epoxy, expand or contract with temperature. The silica changes the rate of thermal expansion of the epoxy to match that of the die and the wires within, reducing stress on the device at the limits of its temperature range.

In addition to the silicon die, one of the two gold bond wires connected to the top of the die is visible through the black epoxy encapsulation.

LM309K Voltage Regulator

The LM309K's silicon chip is a large but relatively simple **INTEGRATED CIRCUIT**, or **IC**: a circuit made with many subcomponents like transistors and resistors, fabricated together on a single piece of silicon.

This IC is a **VOLTAGE REGULATOR**. It takes voltage within some range as an input and provides a stable output at a fixed, lower voltage. The large metal package, type TO-3, helps dissipate the heat

produced as the regulator operates. The three-terminal device has two insulated pins and a third terminal connected to the case.

In closeup, you can see the circuitry on the surface of the silicon die itself. The right two-thirds of the chip is taken up by a large power transistor that regulates the current flowing from the input connection to the output connection.

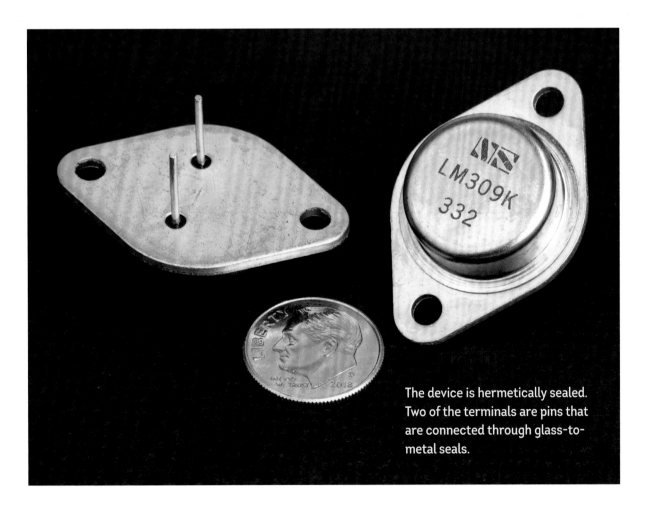

The device is hermetically sealed. Two of the terminals are pins that are connected through glass-to-metal seals.

The input and output bond wires are configured with two wires each in parallel, which doubles their current-carrying capacity.

Dual In-Line Package (DIP) ICs

Integrated circuits can have many more terminals than the two or three on the components that we've looked at thus far.

The **DUAL IN-LINE PACKAGE**, or **DIP**, is a classic package for ICs with a larger number of connecting leads. Two parallel (i.e., *dual in-line*) rows of terminal pins are connected to a stiff metal **LEAD FRAME**, which connects to the central chip through hair-thin bond wires.

Ceramic DIPs are made of two slabs of ceramic on either side of a layer of glass **FRIT**, tiny glass beads that are melted together to form an airtight hermetic seal around the IC and its bond wires. Meanwhile, plastic DIPs are molded, usually with black plastic, directly over the IC, bond wires, and lead frame. A clear plastic DIP gives us a peek inside at the spidery shape of the lead frame, revealing how it's attached to the IC with bond wires.

Light playing across the glass frit creates a rainbow shimmer effect on this vintage ceramic DIP, a Motorola logic chip from 1985.

Each active pin on the package is paired with at least one bond wire that connects to the IC within.

The clear plastic package of this ULN-2232A motion detector IC allows light to reach a square photosensor in the middle of the chip.

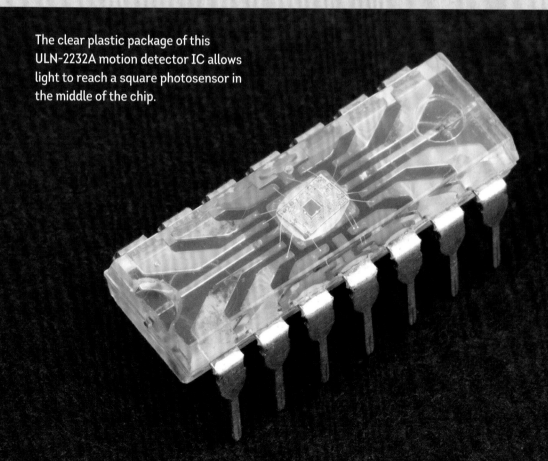

ATmega328 Microcontroller

A **MICROCONTROLLER** is a simple, slow computer on a chip. It's the electronic brain inside many devices: appliances, toys, even flashlights and radios.

The ATmega328 microcontroller is particularly popular with electronics hobbyists and is available in various packages including the 28-pin plastic DIP shown here. An "8-bit" microcontroller, its processing power is in the same general class as that of early-generation home computers like the Apple II.

The black plastic on this device was carefully etched away with concentrated ("fuming") nitric acid, revealing the silicon die within. The individual transistors on a chip like this—there are at least hundreds of thousands—aren't visible at this level of magnification. The etching reveals that the plastic itself is densely filled with silica, just like the 2N3904 on page 72.

This microcontroller is the core component of the Arduino Uno development board.

The actual silicon die is quite small compared to the overall package.

Small Outline Integrated Circuit

Some electronics are still manufactured in dual in-line packages, but smaller and more space efficient surface mount packages, such as the SMALL OUTLINE INTEGRATED CIRCUIT (SOIC) packages shown here, are much more common today. The leads on SOICs are packed more tightly together, spaced apart by only 0.05 inch (1.27 mm), rather than the 0.1-inch (2.54 mm) spacing used for DIPs.

The difference in scale is so pronounced that the silicon chips inside SOIC packages can appear enormous, even though they're actually about the same size as the chips in DIP devices.

One of these SOICs, a color sensor, comes in a clear package that lets us see exactly how the bond wires are connected between the die and the leads.

A 24LC64 serial EEPROM chip stores a small amount of data, equivalent to about 50 text messages, in non-volatile memory.

The silicon die rests on a copper lead frame in the middle of the SOIC. Tiny bond wires connect it to the various leads, but only one of them is visible in this cross section.

Red, green, blue, and transparent color filters help this color sensor perceive the same wavelengths of light as our own eyes.

Thin Quad Flat Pack

Another style of surface-mount chip, the **THIN QUAD FLAT PACK (TQFP)**, has connecting leads on all four sides, instead of on two sides like an SOIC. TQFPs are quite thin, but we'll soon be examining chips that are even thinner.

We removed material from underneath a TQFP so you can see that the actual IC die sits in the middle of the package. The bond wires aren't visible, since they're on the opposite side, but you can see some interesting shapes in the copper lead frame: the profiles are carefully designed so that the epoxy locks the leads in place and prevents them from falling off.

A transparent TQFP—the jewel-like image sensor from an optical mouse—reveals the location and arrangement of the silicon die, lead frame, and bond wires.

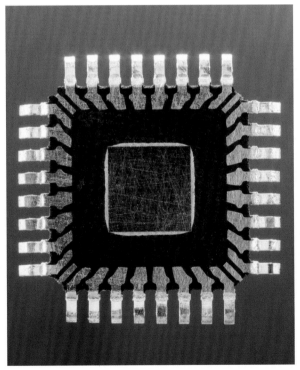

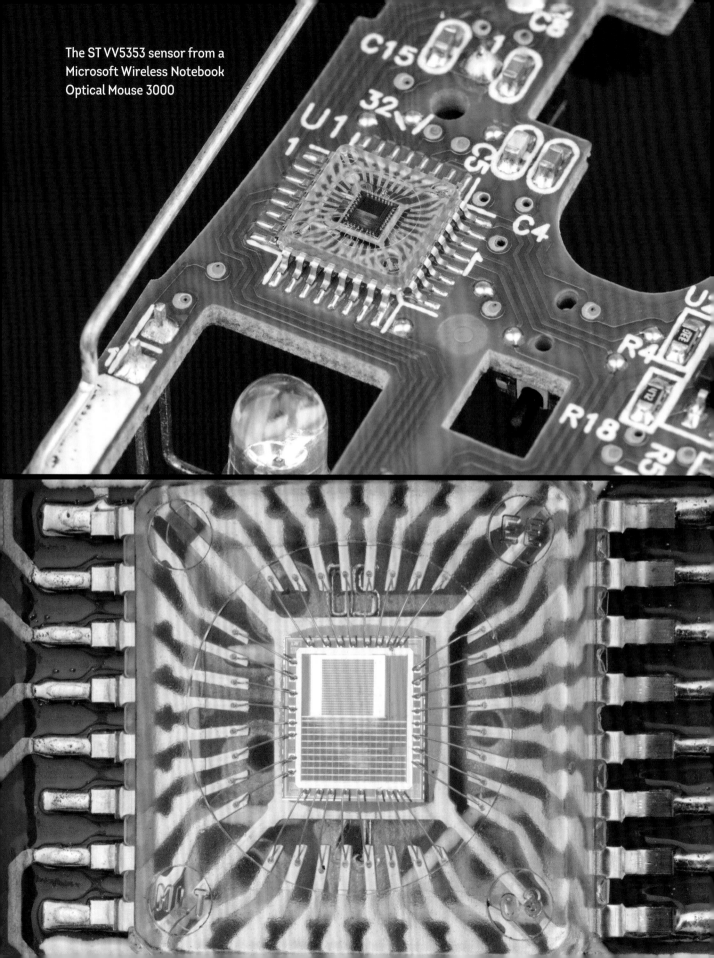

The ST VV5353 sensor from a Microsoft Wireless Notebook Optical Mouse 3000

Ball Grid Array

To save space, many modern chips make connections to a circuit board not through pins or terminals on their sides, but through a grid of tiny balls of solder on the underside of the component. These **BALL GRID ARRAY (BGA)** packages are ubiquitous in modern smartphones, laptops, and other complex, compact electronics.

The solder balls sit on a thin two-layer printed circuit board called a **REDISTRIBUTION LAYER (RDL)** that's embedded into the chip's package. Fine copper traces and I-shaped **VIAS** connect the solder balls on the bottom of the RDL to the bond wires on the top, which make the final connection to the silicon chip itself.

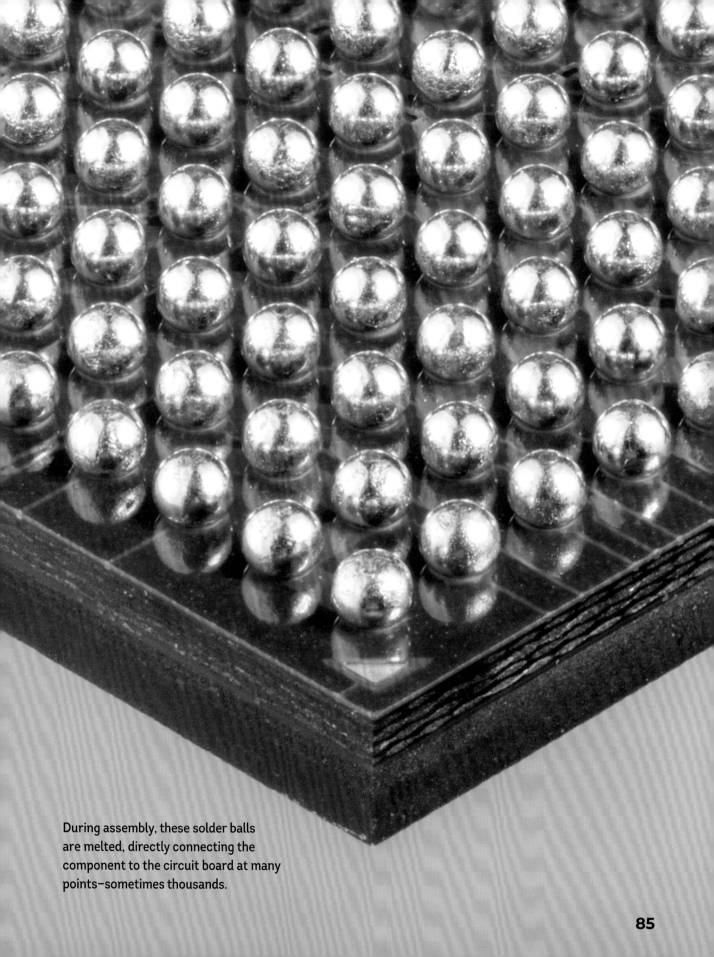

During assembly, these solder balls are melted, directly connecting the component to the circuit board at many points—sometimes thousands.

Microprocessor SoC

A **SYSTEM ON A CHIP**, or SoC, is a high-end microprocessor, integrating a processor and most of the additional functionality, such as graphics support, that would otherwise require separate chips on a computer motherboard. A typical smartphone uses a custom SoC as its main processor, configured with the exact set of features needed for that phone.

The SoC shown here is packaged in a ball grid array for mounting to a circuit board. Inside, the IC die itself is mounted to the redistribution layer using tiny solder bumps applied to the die itself.

Using solder bumps instead of bond wires makes it easier to scale up the number of connections, but a high-density redistribution layer is needed to fan out the connections to the larger BGA. This RDL has 10 layers of copper with laser-drilled passages, called **MICROVIAS**, connecting them.

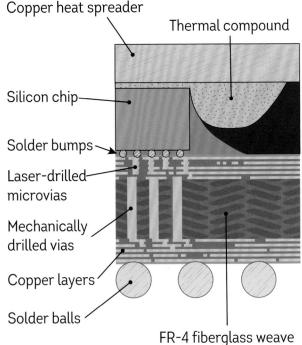

Copper heat spreader

Thermal compound

Silicon chip

Solder bumps

Laser-drilled microvias

Mechanically drilled vias

Copper layers

Solder balls

FR-4 fiberglass weave

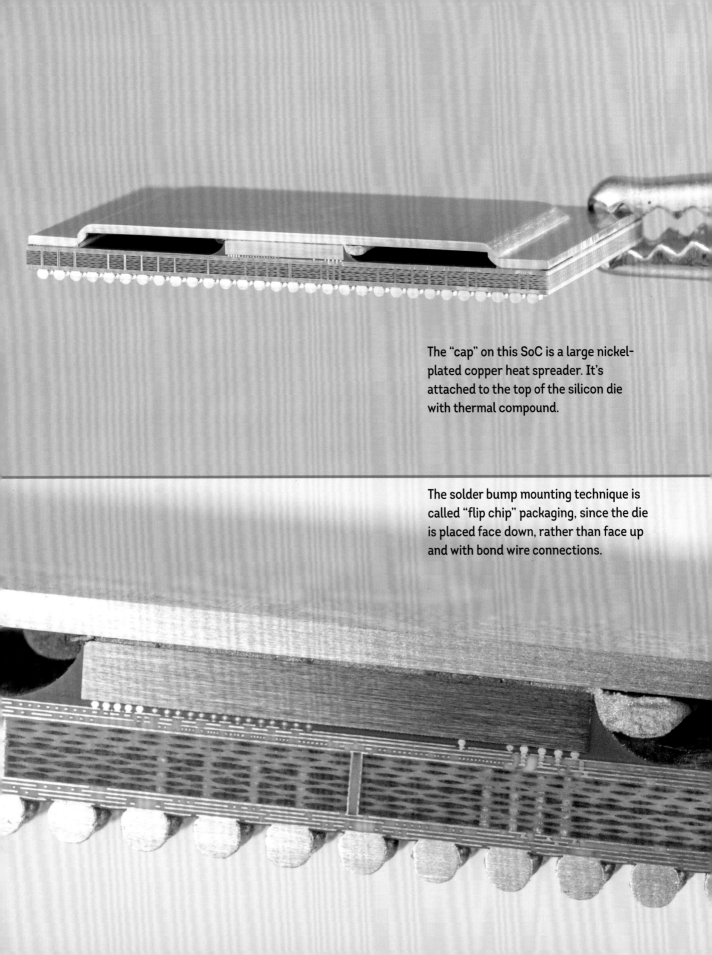

The "cap" on this SoC is a large nickel-plated copper heat spreader. It's attached to the top of the silicon die with thermal compound.

The solder bump mounting technique is called "flip chip" packaging, since the die is placed face down, rather than face up and with bond wire connections.

Through-Hole Red LED

Consistently charming and deceptively simple, LIGHT-EMITTING DIODES, or LEDs, are filled with subtle design details.

The semiconductor die in an LED isn't silicon, but a tailored semiconductor that emits the desired color of light when active. For example, AlGaAs (aluminum gallium arsenide) is typically used to make red LEDs like these.

The odd shapes of the metal connecting leads and fine lines scribed into them help lock the leads in place, securing them in the epoxy molding compound and allowing them to be bent without damaging the fragile LED die. The larger cathode lead is shaped into a reflector cup under the die to direct light forward. A hair-thin bond wire connects the smaller anode lead to the top surface of the die.

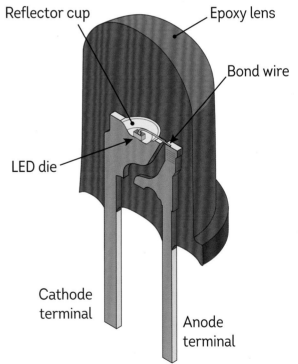

Reflector cup

Epoxy lens

Bond wire

LED die

Cathode terminal

Anode terminal

LEDs are planar devices: only the top surface of the semiconductor die emits light. The dies are cut into neat cubes for ease of handling.

Surface-Mount LED

SURFACE-MOUNT LEDs are just like through-hole LEDs, except for the packaging. Rather than wire leads, a surface-mount LED sits atop a thin circuit board with plated terminals that can be soldered to a larger board. A clear plastic lens is molded directly onto the thin circuit board, encasing and protecting the LED die and its bond wire.

The semiconductor formulation of the LEDs shown here makes them light up green, not red.

This image is a composite of several photos with different exposure times in order to show additional detail.

Red Green Bicolor LED

This **BICOLOR LED** has two leads and two different LED dies inside, connected in parallel with bond wires. A red/green LED lights up red when current flows through it in one direction. If you reverse the voltage, so that current flows in the other direction, it lights up green. With careful circuit design—switching the relative time that current flows each way—the LED can appear to produce red light, green light, yellow light, or anything in between.

LEDs like these are sometimes used as "front panel" indicator lights. Vast arrays of them were used in red/green early-generation LED readerboard displays and gas station signs.

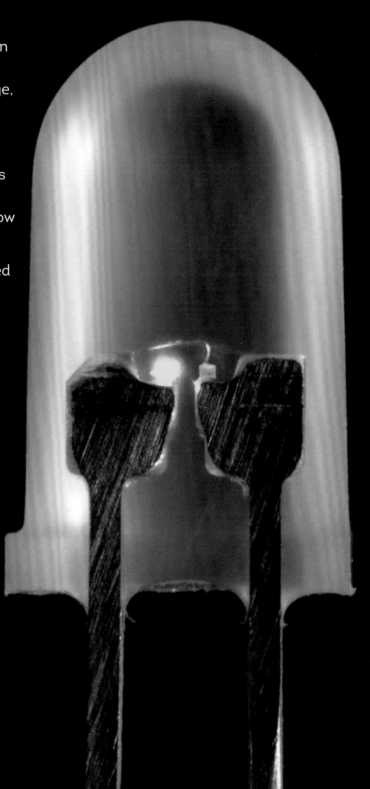

White LED

The device that we call a **WHITE LED** is a chimera: part LED and part clever chemistry. The problem is that true white light contains every color of the rainbow, but an LED can only emit a single color of light, determined by the characteristics of the semiconductor. As a solution, we can fool the human eye into seeing white by mixing red, green, and blue light together.

The white LED's die is located at the bottom of a reflector cup, and it actually emits blue light. The reflector cup is filled with a chemical compound called a **PHOSPHOR**. It absorbs blue light and emits a broad spectrum of colors, tinted somewhat toward red. The phosphor's light combines with the blue from the LED to generate the bright white light that we perceive.

This image, a composite of photos with different exposure levels, reveals a hard-to-capture blue glow around the LED die.

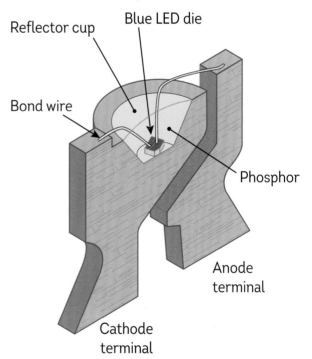

Reflector cup

Blue LED die

Bond wire

Phosphor

Anode terminal

Cathode terminal

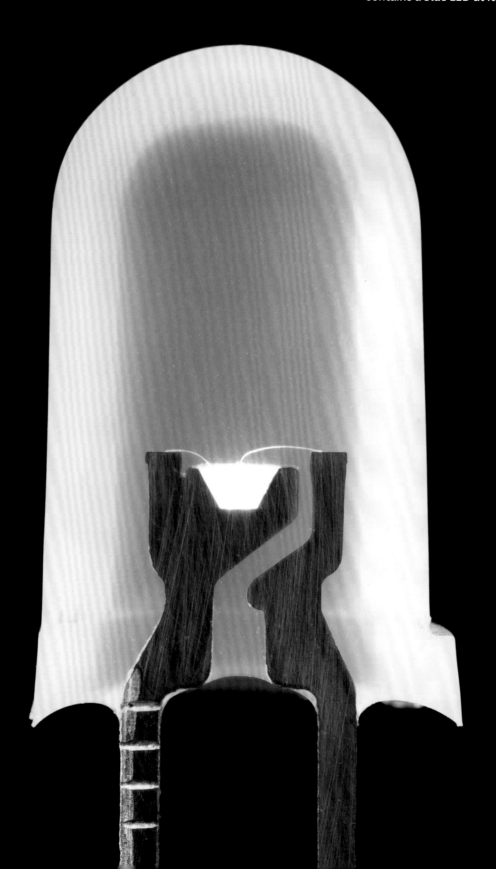

While nothing suggests this from the outside, every white LED contains a blue LED at its core.

Laser Diode

Laser printers are named after the lasers they use to form the image to be printed on the page. The **LASER DIODES** in the picture below are from a modern desktop color laser printer.

Each laser diode is housed in a TO-56 metal "can" package fitted with an antireflection-coated glass window, a heat sink, and a sensitive light detector called a **PHOTODIODE** that measures the amount of laser output.

The laser element itself is the small die with the red face sitting atop the larger silicon photodiode die. Each die is connected to a terminal with a bond wire. A third "common" terminal is connected through the case of the metal can.

When active, the laser die emits a beam horizontally, rather than vertically as an LED does. This type of laser emits in the near infrared, just beyond the reddest red that human eyes can see.

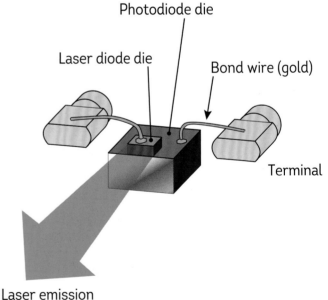

Photodiode die

Laser diode die

Bond wire (gold)

Terminal

Laser emission

The laser diode emits light not only out the front, but also out the back. The angled surface on the rear face of the package reduces unwanted direct reflections.

Optocoupler

An **OPTOCOUPLER** converts an electrical signal to light and back again. It provides electrical isolation, much like a transformer but using light instead of a magnetic field.

An LED, which converts an electrical signal to light, is mounted at the top, facing down toward a **PHOTOTRANSISTOR**, a light sensor that converts the LED's light back to an electrical signal. The LED die is protected by a bead of clear silicone. The device is molded with a translucent plastic to allow light to pass within the component, and black plastic over that to prevent interference from external light.

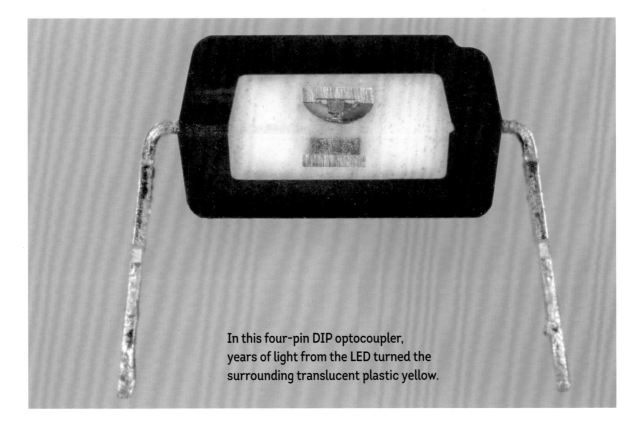

In this four-pin DIP optocoupler, years of light from the LED turned the surrounding translucent plastic yellow.

Optical Tilt Sensor

OPTICAL TILT SENSORS were used in early-generation digital cameras to determine camera orientation when taking a picture. They contain an infrared LED pointed at two phototransistors.

A small metal ball, free to roll about, sits between the LED and the sensors. When upright, there's a clear path over the ball from the LED to both sensors. When the device is tilted left or right, the ball rolls, blocking the light from reaching one phototransistor or the other.

The LED, in transparent pink epoxy, emits light toward a die with two phototransistors, encased in black plastic that's transparent to infrared.

Optical Encoders

In a modern computer mouse, a low-resolution optical sensor (like the one we saw on page 83 in a thin quad flat pack) measures the mouse's change in position as you move it around. In old-style ball mice, two **OPTICAL ENCODERS** sense the motion of a physical ball that rolls when the mouse moves.

Optical encoders work like an advanced version of the optical tilt sensor. An infrared LED shines light through an **ENCODER WHEEL**, a wheel with slits that alternately block or allow the light to pass. Two phototransistors positioned on the other side of the wheel detect the light as the wheel rotates. Circuitry in the mouse decodes the phototransistors' output signals to calculate how far and in which direction to move the on-screen cursor.

Today, ball mice are obsolete, but mice with scroll wheels still use optical encoders to detect the rotation of the wheel.

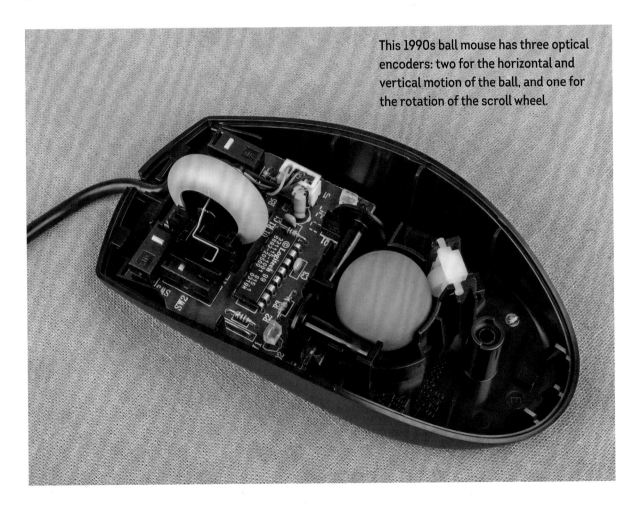

This 1990s ball mouse has three optical encoders: two for the horizontal and vertical motion of the ball, and one for the rotation of the scroll wheel.

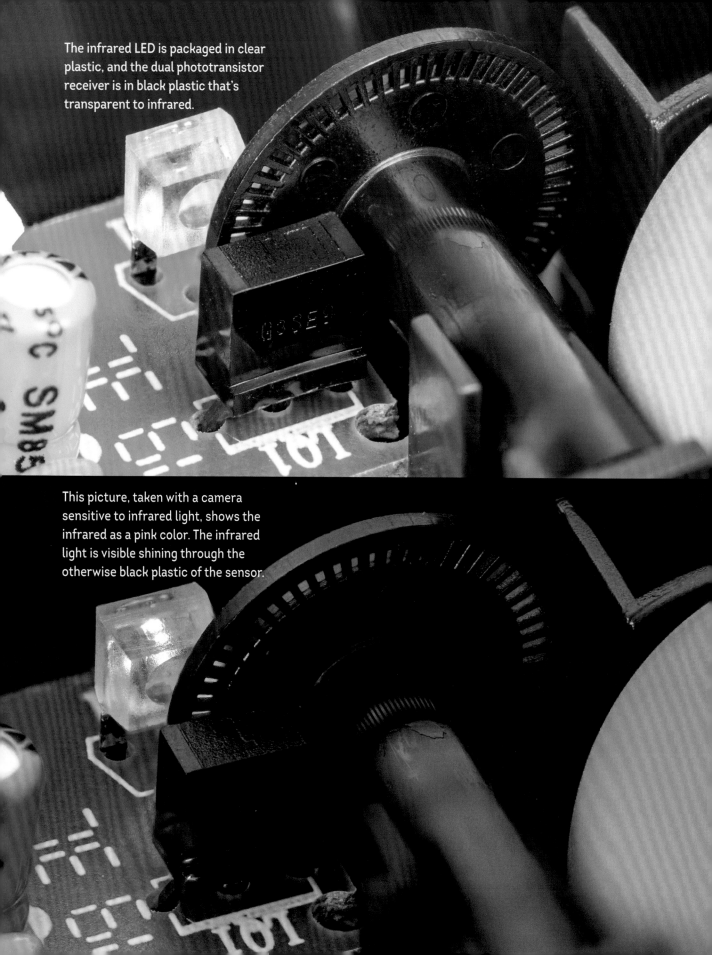

The infrared LED is packaged in clear plastic, and the dual phototransistor receiver is in black plastic that's transparent to infrared.

This picture, taken with a camera sensitive to infrared light, shows the infrared as a pink color. The infrared light is visible shining through the otherwise black plastic of the sensor.

Ambient Light Sensor

A tiny **AMBIENT LIGHT SENSOR**, only 1 mm wide, sits between the camera and LED flash on a smartphone. It measures the amount and character of light, so the phone can sense and compensate for the color temperature of a photographed scene. The sensor also allows the phone to adjust the screen's display color and brightness depending on its surroundings.

The device has a six-pin interface, a minuscule 2×3 ball grid array. Its clear package shows that the die is almost the entire size of the overall device. The sensor portion of the die features 25 squares with different optical filters for sensitivity to different colors: red, green, blue, and also invisible infrared and ultraviolet light.

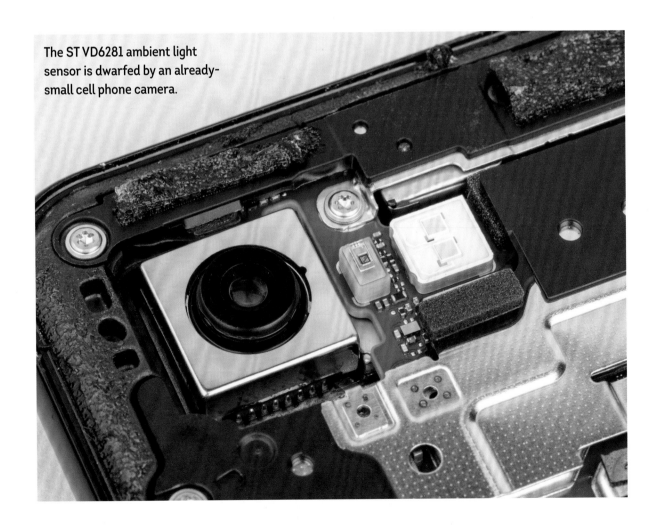

The ST VD6281 ambient light sensor is dwarfed by an already-small cell phone camera.

The sensor sits on a circuit board riser that raises it to about the same height as the flash on the back surface of the phone. This gives the sensor the widest possible field of view.

CMOS Image Sensor

All semiconductor devices are inherently sensitive to light. Put an array of them on a chip and you get an **IMAGE SENSOR** that can convert a two-dimensional image into an electrical signal. Such a chip forms the heart of a digital camera.

The image sensor shown here sees in black-and-white, but an optical filter with a red, green, and blue checkerboard pattern is applied to the image sensing matrix, allowing it to perceive color.

Complex circuitry visible at the top of the die generates the control signals that drive the array, amplifies the small signals from the image sensor, and converts them into digital data that can be processed, stored, and uploaded to your social media accounts.

The designation CMOS (complementary metal oxide semiconductor) refers to the specific fabrication process used to manufacture the device.

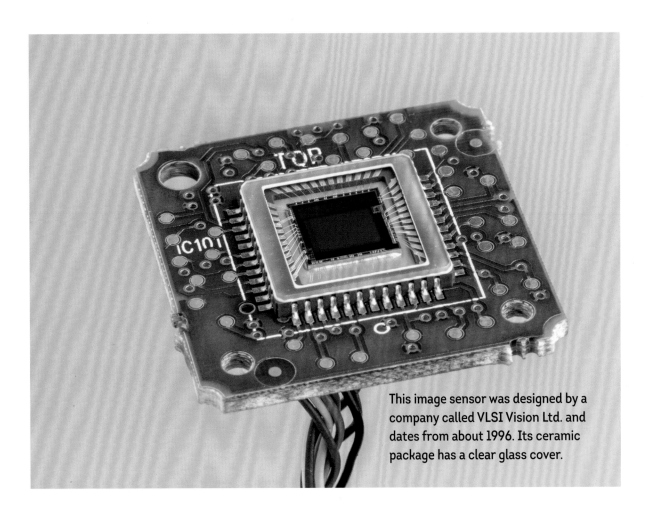

This image sensor was designed by a company called VLSI Vision Ltd. and dates from about 1996. Its ceramic package has a clear glass cover.

Electromechanics

Most of the devices that we've looked at so far have no moving parts. But many important components straddle the worlds of electronics and mechanics. Switches, motors, speakers, electromagnetic relays, hard disk drives, and smartphone cameras might seem unrelated to each other, but a common thread ties—or *wires*—them together.

Toggle Switch

With the flick of a finger, a **TOGGLE SWITCH** flips back and forth between two positions.

Inside, the mechanism is surprisingly simple. A metal bar teeter-totters between two positions, connecting a common center lead to one of two internal contacts. Electrical current is therefore routed to one of two possible paths.

The plastic finger that presses down on the metal bar is spring-loaded so that it

snaps into either position and provides consistent pressure between the bar and the contact. Being plastic, it keeps the lever that you touch electrically insulated from any voltage that may be on the terminals.

Similar toggle switches may add an "off" position in the middle, or additional **POLES**—independent sets of contacts that are switched in parallel by the same lever.

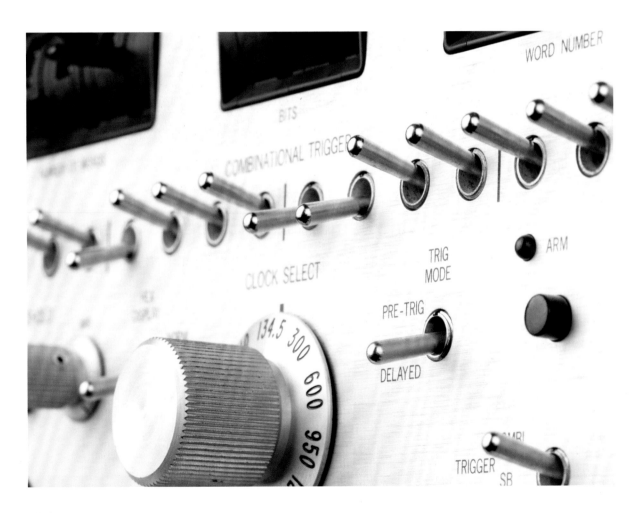

The upper section of the switch is threaded for mounting to a panel. The pin through the threaded section is the pivot point for the lever.

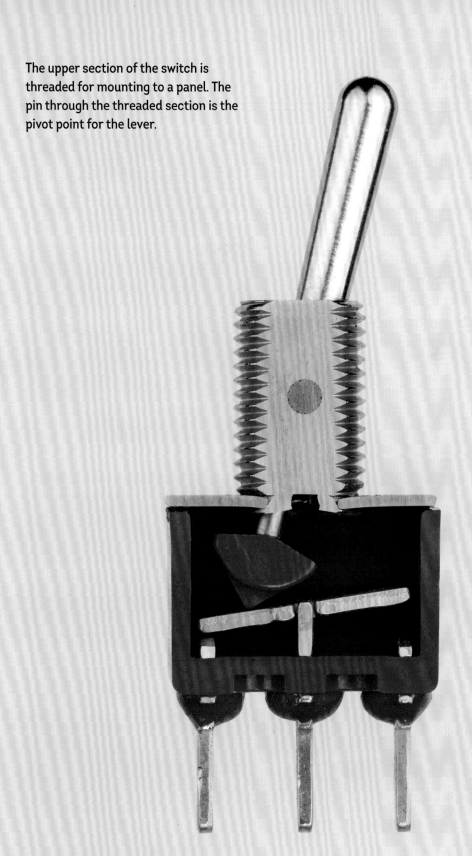

Slide Switch

The two-position **SLIDE SWITCH** shown here has small ridges on its handle, making it easy to grip with your fingertip and move it from one position to the other.

Inside, the handle slides a metal contact plate back and forth, completing a circuit between the middle terminal and one of the two outer terminals. Larger and more complex slide switches may add additional terminals and positions for the sliding handle.

A compression spring sits between the handle and the metal contact plate, pressing the plate against the terminals as it slides.

Pushbutton Switch

This basic **PUSHBUTTON SWITCH** is often found on the front panels of hobby electronics projects, though it isn't typically used in commercial products. Still, its basic principle of operation applies to other, more common types of pushbutton switches.

Pressing down the spring-loaded button moves a metal washer down against two contacts, connecting them and completing the circuit. When you let go, the spring pushes the washer back up, breaking the circuit. This is said to be a "normally open" switch, since it gives an open circuit unless the button is pressed.

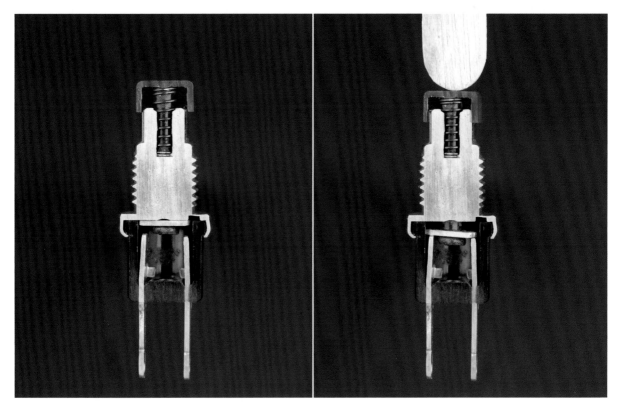

DIP Switch

You may have come across and set DIP SWITCHES yourself; they're common in alarm control panels, industrial equipment, household heating controls, and some older computers. They're named for the familiar dual in-line arrangement of their two rows of terminals. Each pair of terminals has its own switching mechanism.

There are several different styles of DIP switches, based on miniaturized slide switches or different toggle and rocker switch mechanisms. The one shown here uses a simple rocker type mechanism.

Inside each switch element are a white plastic rocker, a spring-loaded metal ball, and two contacts. Changing the rocker position moves the ball so that it's either tucked away to the side or connecting the two contacts together.

DIP switches set the configuration of an Apple Super Serial Card II inside an Apple IIe computer.

The gold-plated metal ball is just under 1.5 mm in diameter. An eight-position DIP switch contains eight balls held in place by eight springs.

Tactile Switch

TACTILE SWITCHES come in many sizes and are widely used as responsive buttons on electronics and appliances. They're often hidden behind larger customized buttons, like the eject button on an optical disc drive or the front panel buttons on a home entertainment system.

Pushing down on the button causes a thin metal dome inside the switch to collapse and complete an electrical circuit. As soon as you let go, the dome snaps back into shape, breaking the circuit and stopping the flow of current. The springy metal dome creates a satisfying clicking sound and a quintessential tactile feel.

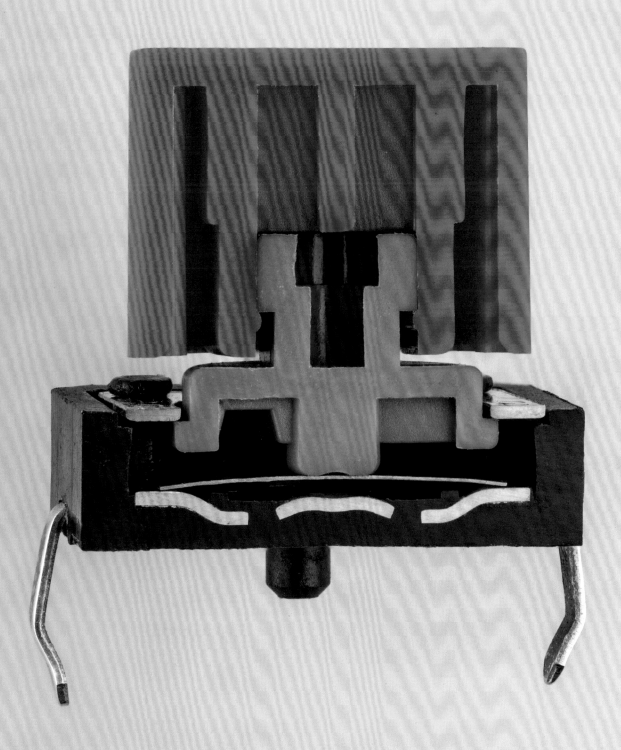

This tactile switch has a large red button cap. Low-profile versions are ubiquitous in electronics.

Microswitch

MICROSWITCHES provide the electrical function and clicky feel to the buttons on computer mice. They're remarkably reliable switches, designed to work for millions of cycles.

Inside, two stamped metal springs—one straight and flat, the other curved—play off each other to create a consistent snap action when a plunger is depressed past a certain tripping point. Releasing the plunger snaps the switch back to the other position. The snap action moves a common electrical contact between two fixed contacts attached to the terminals of the device.

Beyond computer mice, you can find this type of microswitch in any number of industrial and automation applications. Common examples include limit switches on 2D and 3D printers.

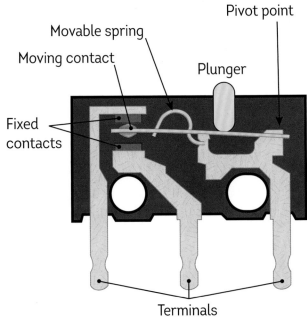

Pivot point

Movable spring

Moving contact

Plunger

Fixed contacts

Terminals

Electromagnetic Relay

ELECTROMAGNETIC RELAYS are switches that are actuated by electrical signals, rather than by a button or lever. They provide a robust, low-cost method of switching substantial electrical power, and are used in appliances, cars, elevators, industrial equipment, and even traffic lights.

The heart of the relay is a SOLENOID, a type of inductor specifically designed to be used as an electromagnet. When current passes through the solenoid's coil of wire, it creates a magnetic field that attracts a hinged iron plate, moving a set of switch contacts from one position to another. When the solenoid shuts off, a spring retracts the iron plate, pulling it and the switch contacts back to their initial positions. Thus, the device uses electricity to relay electricity.

This relay has four poles: it uses one solenoid to simultaneously actuate four switches that can control four independent signals.

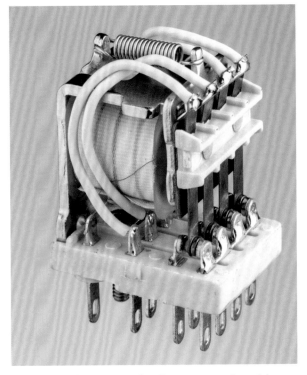

The solenoid is wrapped in fine copper wire with an outer layer of cloth tape. Thick rubber-insulated wires connect to the center terminal of each switch.

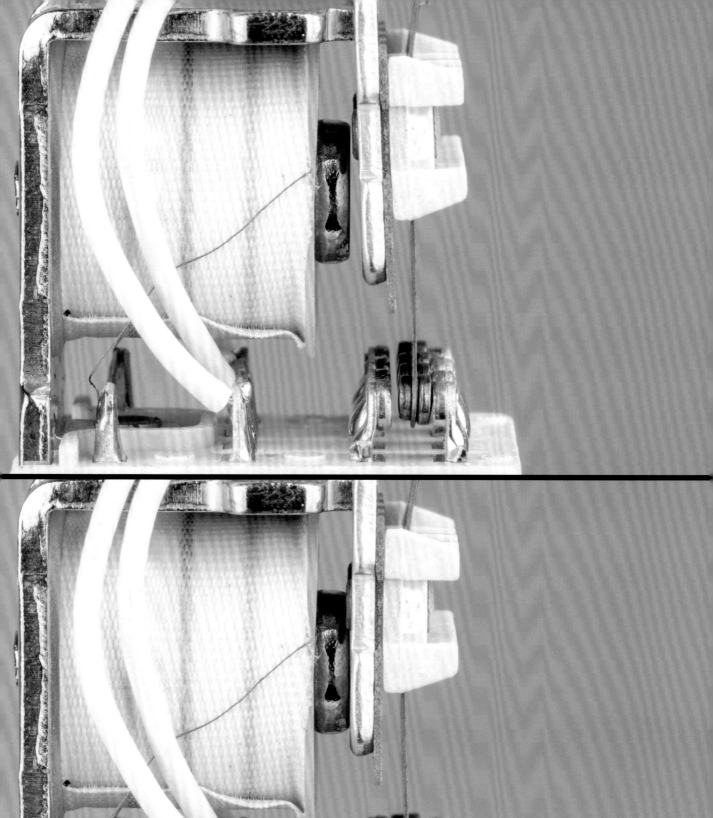

Thermal Switch

Simple devices that need to regulate a temperature do so using a **THERMAL SWITCH**, an electrical switch that opens or closes at a given temperature. For example, a thermal switch in a coffee maker can turn a heater on every time the warming plate's temperature falls below a given setpoint.

The active element of a thermal switch is a **BIMETALLIC STRIP**, a welded sandwich of two dissimilar metals with different rates of thermal expansion. In the switch shown here, the bimetallic element is a thin disk that changes shape as it heats and cools.

At room temperature, the disk is flat. It pushes up a small ceramic rod, pressing two electrical contacts together to connect them. When the temperature rises above a fixed setpoint, the disk becomes concave, bowing downward so that the ceramic rod releases the contacts and disconnects the circuit.

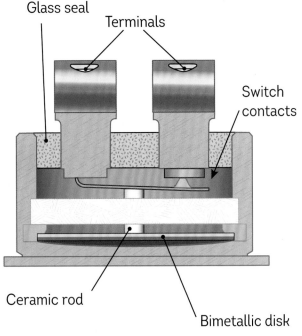

Glass seal

Terminals

Switch contacts

Ceramic rod

Bimetallic disk

This thermal switch has been cut in half to show its mechanism. Normally, a switch like this is sealed to prevent dust from getting inside.

Brushed DC Motor

This miniature "pager" motor is about the same diameter as a pencil. It's one of several types commonly used as the vibrating motor inside a phone.

Current flowing through the copper windings inside the motor generates a magnetic field that pushes against another field from a permanent magnet. The permanent magnet is called the **STATOR** since it's fixed in place. The windings are attached to a shaft—the

ROTOR—that begins to turn due to magnetic attraction and repulsion.

Metal fingers called **BRUSHES** conduct electrical current into the rotating copper windings even as they turn. These brushes also act as a **COMMUTATOR**, reversing the polarity of the current through the copper windings every half turn. Otherwise, the rotor would simply align with the stator magnet and stop turning.

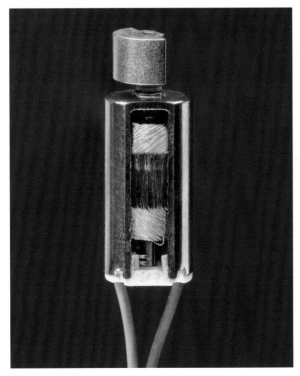

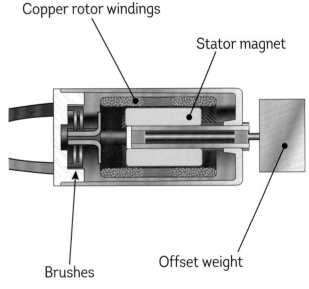

Copper rotor windings

Stator magnet

Brushes

Offset weight

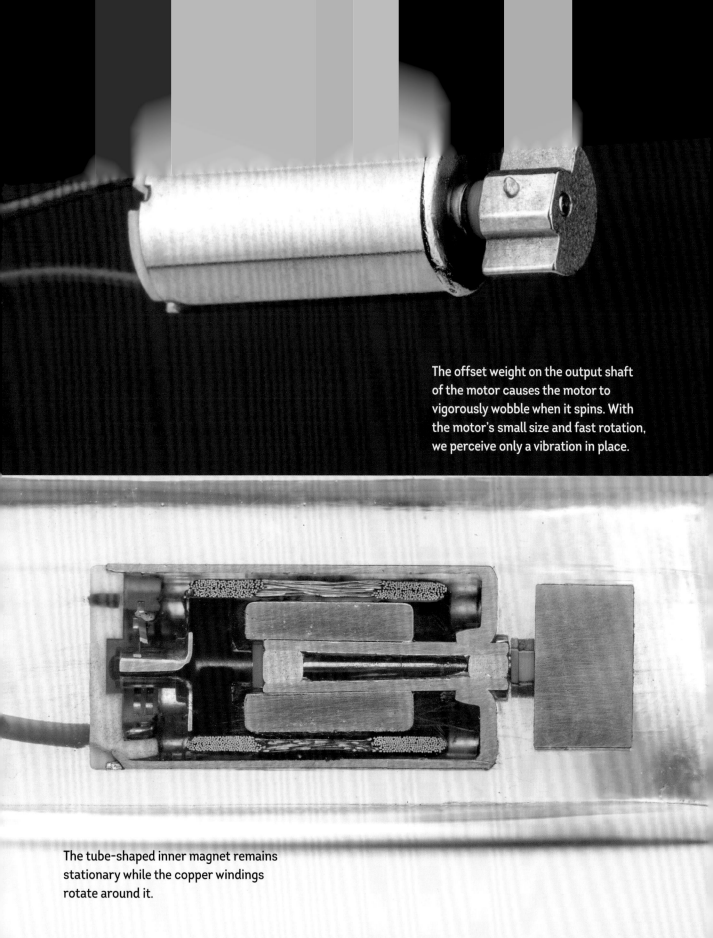

The offset weight on the output shaft of the motor causes the motor to vigorously wobble when it spins. With the motor's small size and fast rotation, we perceive only a vibration in place.

The tube-shaped inner magnet remains stationary while the copper windings rotate around it.

Stepper Motor

While many motors are designed to rotate continuously, **STEPPER MOTORS** are optimized for starting and stopping quickly in precise rotational increments called "steps."

Steppers are **BRUSHLESS** motors, meaning the copper windings are part of the nonmoving stator while the permanent magnet is part of the moving rotor. The stepper motor shown here is a common type used in desktop 3D printers. It has eight copper windings: two opposing pairs of four, wrapped around a stator form constructed of laminated iron sheets. The rotor has its own stacked iron layers that act as the pole pieces for a high-strength permanent magnet.

Toothed structures on both the rotor and stator determine the motor's step size, or resolution. This motor can move in 200 discrete steps per revolution.

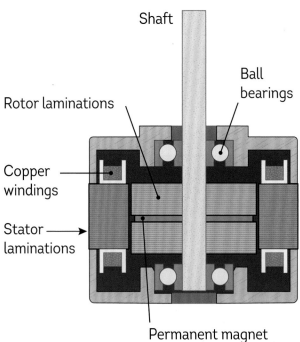

Shaft

Ball bearings

Rotor laminations

Copper windings

Stator laminations

Permanent magnet

When assembled, the rotor fits neatly inside the stator, with only a thin gap between the two.

In cross section, we can see not only the copper windings but also the ball bearings and the laminated rotor and stator.

Magnetic Buzzer

Many different kinds of equipment use MAGNETIC BUZZERS to make all sorts of noises: alarm sounds, informational beeps, and even simple tunes to alert you that, for example, your rice has finished cooking. A PC motherboard uses a magnetic buzzer to alert you when it has a low-level failure.

The insides of this drab component are unexpectedly vibrant. The most eye-catching part is the small solenoid of magnet wire wrapped around an iron core. When current is applied to the two connecting leads, the copper wire generates a magnetic field in its core. That field combines with the magnetic field from the ring-shaped magnet around the outside of the coil, and pushes against the metal diaphragm in the middle. When driven with an alternating current signal, the diaphragm vibrates at the frequency of the input signal, producing a tone.

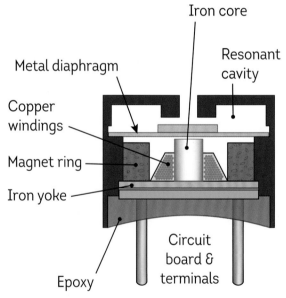

Metal diaphragm

Copper windings

Magnet ring

Iron yoke

Iron core

Resonant cavity

Circuit board & terminals

Epoxy

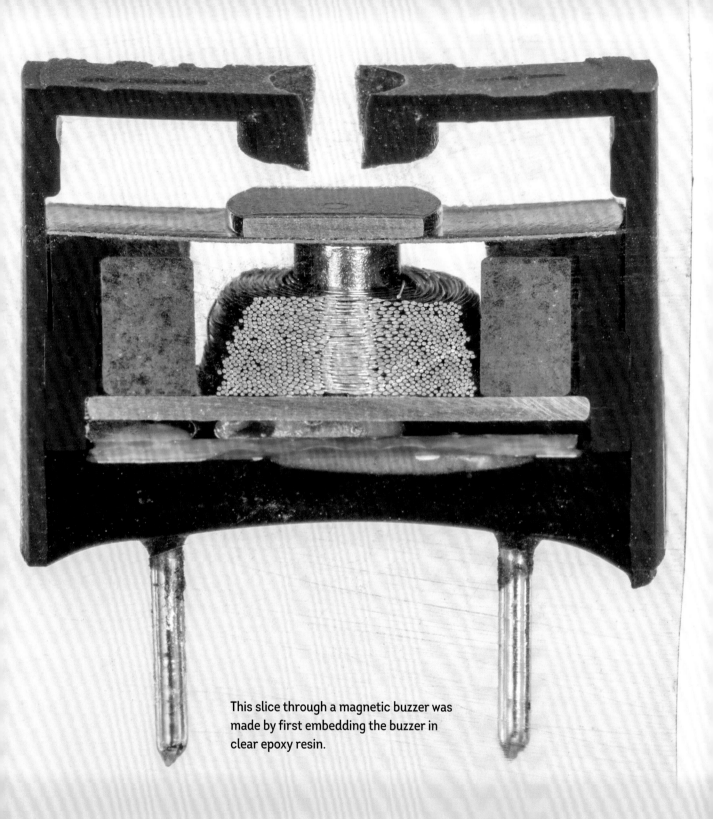

This slice through a magnetic buzzer was made by first embedding the buzzer in clear epoxy resin.

Speaker

A **SPEAKER** converts an electrical signal into vibrations of air that we perceive as sound.

Inside the speaker, a large permanent magnet sits at the center. A small coil of wire called a **VOICE COIL** is wound around a paper cylinder that fits inside a circular groove in the magnet. When the voice coil is driven with current in either direction, it generates a magnetic field that pushes against the field of the permanent magnet. The paper

cylinder moves up or down in response. An amber-colored suspension acts as a spring to return the cylinder to the neutral position when no current is present.

The paper cylinder connects to a black speaker cone made of molded paper that provides a good surface area for pushing against the surrounding air. The cylinder's movements drive the vibrations of the speaker cone, generating the sound waves that we hear.

A small speaker mounted to the case of a vintage Apple IIc computer

This thin slice through a speaker was cast in transparent resin.

In a close-up view, you can see that the red-lacquered copper magnet wire and attached paper cylinder are precisely positioned within the groove of the permanent magnet, free to move up and down.

Smartphone Camera

One of the most mechanically complex parts of a smartphone is the camera assembly. In addition to an image sensor with several megapixels of resolution, it contains a multi-element lens, an infrared blocking filter, and an auto-focus mechanism.

And all of this fits within about a cubic centimeter of volume.

What does a smartphone camera have to do with the electromechanical devices we've been looking at so far? The answer is that the auto-focus mechanism uses a voice coil motor to precisely position the lens with respect to the sensor, just as though the lens were the paper cone in a speaker.

One of this book's authors used this same type of Nexus 5X smartphone camera assembly to take his first cross-section photos and post them on his Twitter account.

The optical assembly includes six precision molded plastic lenses with aspherical profiles. The focusing mechanism changes the distance between the lens assembly and the sensor.

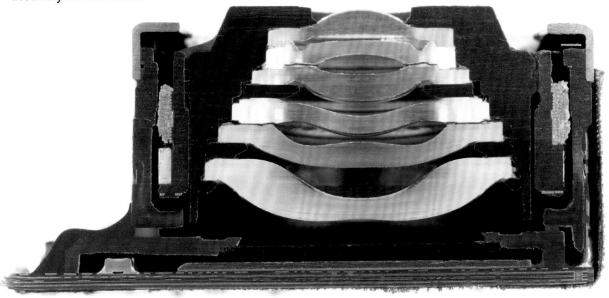

Inside the Camera Module

The camera uses a copper voice coil around the lens assembly to precisely position it with respect to nearby fixed magnets. Varying the amount of current causes greater or lesser displacement.

When you take a picture using your smartphone and tap on an object to get it into focus, a complex interaction between software and servo circuitry determines the precise magnetic field required to position the lenses in just the right place relative to the image sensor so your image is perfectly in focus.

Beneath the lens assembly is a glass filter for blocking infrared light, and underneath that is the image sensor, sitting on a multilayer circuit board. The image sensor is connected to the board with an array of bond wires.

Part of the infrared cutoff filter—which preferentially reflects red light—has been broken off, revealing the main image sensor below.

Rotary Voice Coil Motor

Hard drives use a high-performance **ROTARY VOICE COIL MOTOR** to move their read/write heads quickly into different positions.

The operating principle is the same as that for the voice coil in a speaker. However, the magnets and coils are arranged to cause the coil to rotate about a pivot point rather than move in a straight line.

Substantial drive power and complex closed-loop servo circuitry allow the hard drive to precisely reposition its heads in only a few milliseconds.

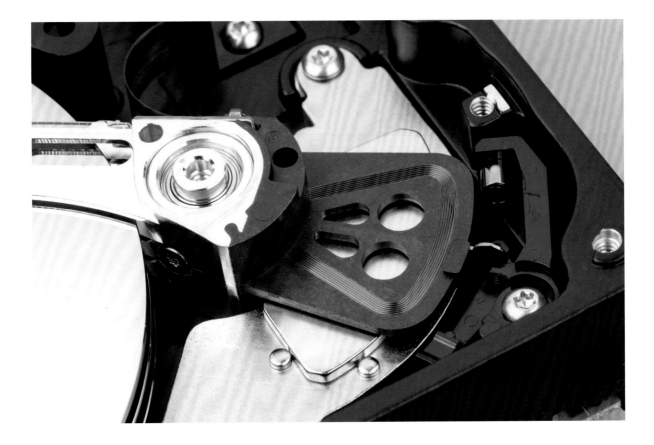

Optical Drive Focusing Motors

This laser assembly from a laptop DVD drive uses a clever two-axis voice coil motor system to position the lens. Two sets of coils and magnets move the lens up and down for focus and side to side for tracking. A DVD's data is encoded in a spiral of digital 1s and 0s. Coarse tracking along the spiral is provided by a DC motor that moves the whole laser assembly along a linear track. For fine adjustments, the tracking coil shifts the lens from side to side.

The lens assembly is suspended in place by long, thin spring wires, which also form the electrical connections to the coils.

Electret Microphone

An **ELECTRET MICROPHONE** is an inexpensive device frequently used as the microphone in consumer electronics applications like phone headsets. It's named after the rather strange material that forms its diaphragm.

An **ELECTRET** has a permanent electric charge stored inside the material itself, a bit like how a magnet has a permanent amount of magnetism. The microphone's electret diaphragm and the **PICKUP PLATE** form a simple capacitor, permanently charged by the electret. When a sound wave impinges on the electret diaphragm, it changes its distance to the pickup plate, which varies the capacitance, and thus generates an electrical signal. The pickup plate is connected to a built-in transistor that amplifies the signal and sends it out through the terminals.

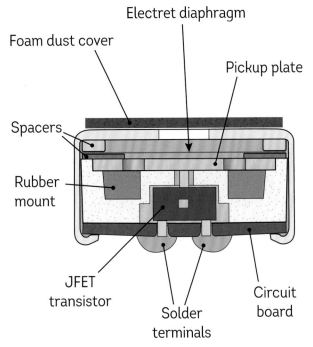

Electret diaphragm

Foam dust cover

Pickup plate

Spacers

Rubber mount

JFET transistor

Solder terminals

Circuit board

This cross section was made by embedding the microphone in clear resin before cutting it. The cut goes through the die and bond wires of the transistor in the black package.

Cables and Connectors

Cables and connectors link our devices to the world around them. They transport electrons, carrying power through our homes, delivering the internet to our computers, streaming video to our screens, and bringing music to our ears. They range from simple strands of wire to incredibly complex feats of precision manufacturing.

Solid and Stranded Wire

Wires are everywhere. They're at the bottom of the ocean and in distant space probes. They move electrical signals through our walls, across continents, and sometimes even within our bodies.

There are two basic types of wire: SOLID and STRANDED. Solid wire has a single filament of metal, while stranded wire consists of multiple smaller wires nestled together. Seven-strand wire is common because the overall shape is roughly circular. For the same reason, you can also find wire with 19, 37, or even 61 strands. Stranded wire is more flexible, while solid tends to hold a shape but break if it's bent too many times.

Wire can be made with any number of different metals, but copper is the most commonly used in small electronics. To prevent electrical short circuits, wire is often covered with insulation, such as varnish, PVC plastic, or even cloth.

AC Power Cable

A bundle of wires is called a cable. Shown here is the type of power cable that might come with a desktop computer sold in the USA. It has a grounded three-prong plug known as type NEMA 5-15, rated for 15 amperes of current.

Inside the cable are three stranded copper wires, sheathed in a molded black outer jacket. The green wire is the ground conductor, and the black and white pair carry the single-phase 120 volt, 60 Hz alternating current (AC). The black wire is "hot," at roughly 120 V with respect to ground, while the white wire is "neutral," at a voltage near to ground.

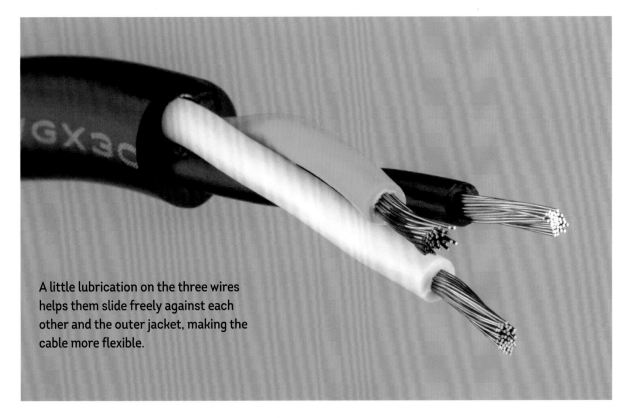

A little lubrication on the three wires helps them slide freely against each other and the outer jacket, making the cable more flexible.

IDC Ribbon Cable

RIBBON CABLES—sometimes colorful—were once very common in computers, and are still used in industrial equipment and hobby electronics. They're shaped like long, flat ribbons, with many individual wires side by side.

The plug type shown here is called IDC, for INSULATION DISPLACEMENT CONNECTOR. It works by forcing each insulated wire between two wedge-shaped metal blades. The blades pierce the insulation and grip the copper wire tightly, making a solid electrical connection. Gold-plated fingers inside the connector make contact with the metal "header" pins that the connector plugs into.

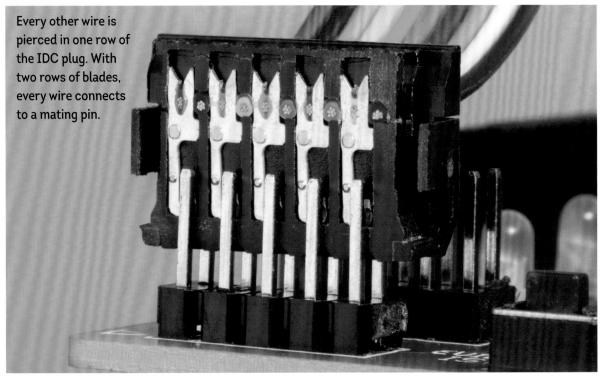

Every other wire is pierced in one row of the IDC plug. With two rows of blades, every wire connects to a mating pin.

Modular Telephone Cable

Older, wired telephones sometimes use this type of flat cable to hook up to an analog telephone line, or landline. The clear connector on the end is a modular plug, part of an RJ25 (registered jack) interface that can work with up to three phone lines. The plug connects to the individual wires through insulation displacement, much like in the ribbon cable connector.

The center two wires, colored green and red, carry the first analog telephone line. The other telephone lines are carried on the remaining two pairs, yellow/black and blue/white. Sometimes the additional wire pairs are instead used to supply low-voltage electrical power to the telephone.

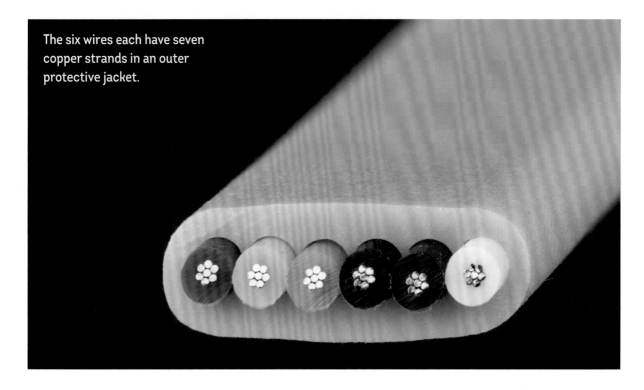

The six wires each have seven copper strands in an outer protective jacket.

DIP Sockets

DUAL IN-LINE PACKAGE (DIP) SOCKETS allow an integrated circuit to be plugged into a board and easily removed, without the need for soldering equipment. Instead, the socket itself has pins that are soldered to the board.

DUAL-WIPE sockets have flat metal springs that push against either side of each pin of the IC. These are inexpensive to produce, as only a single stamped and formed piece of metal is required to connect with each pin.

A **MACHINED-PIN** socket is more complex. The metal sockets are individually machined to the correct shape using a specialized lathe. To grip the pins of the IC, a minuscule stamped and formed set of spring fingers press-fits into each socket.

The socket's pins are typically tin plated, but high-end sockets often have a gold plating to prevent corrosion that could break the electrical connection.

Dual-wipe sockets have springy contacts that grip each pin of the integrated circuit package.

Machined-pin sockets are manufactured with greater precision than dual-wipes. They use press-fit spring connections to grip each IC pin.

Barrel Plug and Jack

BARREL PLUGS and JACKS are commonly found on electronics that use a plug-in AC adapter.

The plug has an outer metal barrel and a center socket. The polarity depends on the specific piece of equipment: sometimes the center socket is the positive terminal and the outer barrel negative, but there's no consistent standard. The jack has a center pin that fits into the plug's socket, and an outer contact that touches the plug's barrel.

Embedded within the jack is a simple switch that some devices use to change from battery power to external power. Inserting the plug into the jack automatically opens the switch, disconnecting the internal battery.

Barrel jack on a circuit board

Barrel plug on a power cable

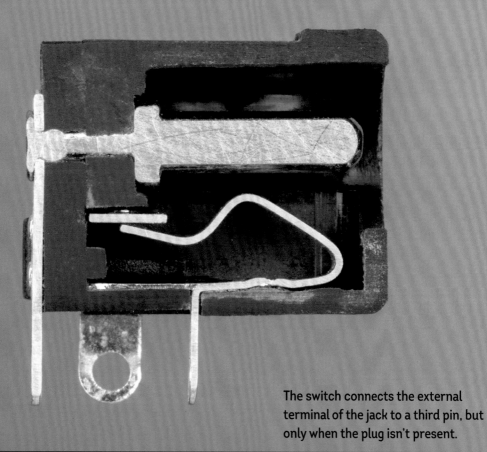

The switch connects the external terminal of the jack to a third pin, but only when the plug isn't present.

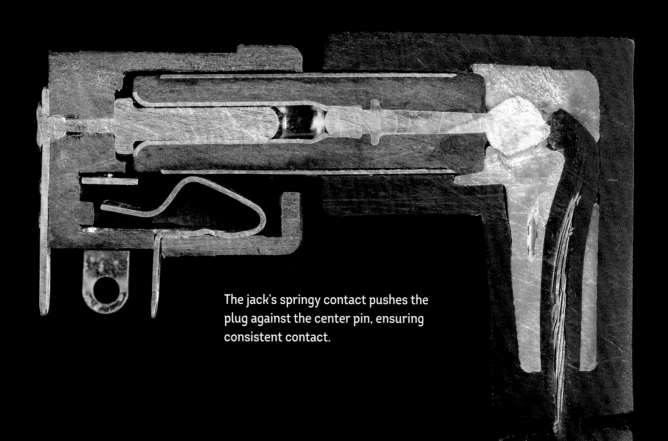

The jack's springy contact pushes the plug against the center pin, ensuring consistent contact.

Quarter-Inch Audio Plug and Jack

The **QUARTER-INCH (6.35 mm) AUDIO PLUG**, sometimes called a **PHONE PLUG**, was one of the first connectors ever invented. Originally designed for telephone switchboards, it has remained relatively unchanged since the 1890s. Operators would connect your call by taking your phone line, which had such a plug, and plugging it into a jack that represented the destination of your call.

The jack has a spring finger that locks into the groove at the tip of the plug, retaining it so it can't readily fall out. Like the barrel jack, quarter-inch jacks have a switch that can detect when the plug has been inserted.

While phone systems don't use these connectors any longer, they're still standard for musical instruments such as electric guitars and synthesizers.

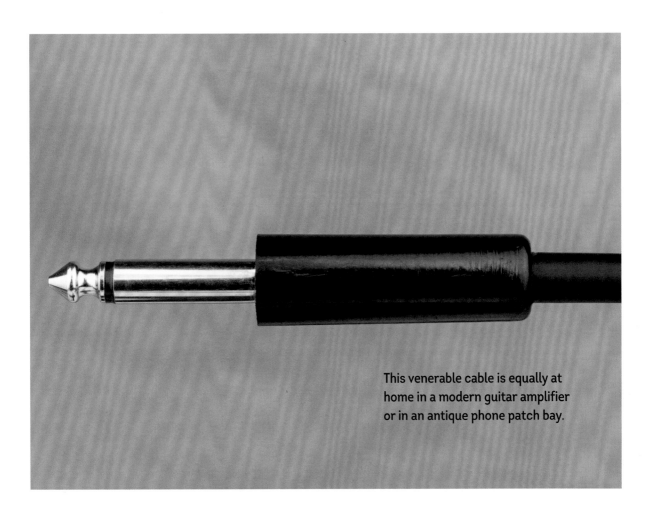

This venerable cable is equally at home in a modern guitar amplifier or in an antique phone patch bay.

The tip of the plug is connected to the center wire of the cable while the outer sleeve is connected to the cable's outer shielding.

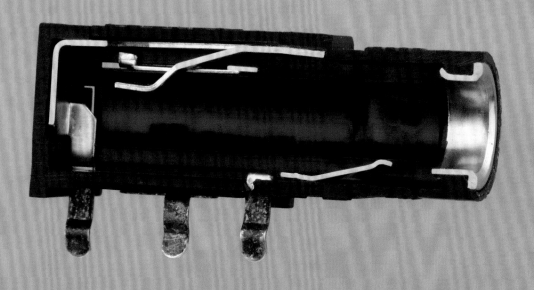

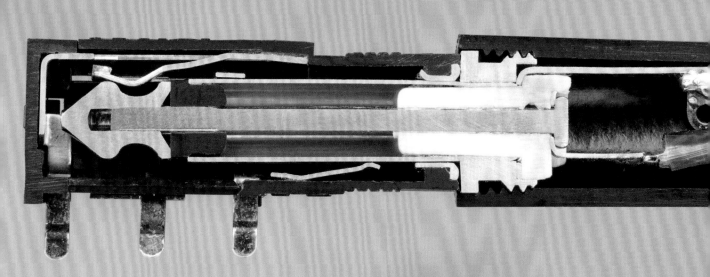

3.5 mm Audio Connector

This familiar audio connector is a miniature version of the quarter-inch audio plug. It's commonly known as a **HEADPHONE PLUG** (though it's also used for other audio signals) or as an **EIGHTH-INCH PLUG** (though that's only an approximation of its size). These connectors are being replaced by USB-C and Bluetooth on smartphones, but they're still the simplest way to get audio in and out of a device.

Other than size, the major difference between this plug and the quarter-inch audio plug on page 146 is that it has three terminals, called the *tip*, *ring*, and *sleeve*, in order to support two-channel stereo audio.

The 3.5 mm jack has two tiny switches inside for disconnecting any internal speakers when you plug your headphones into a device. Some computers use the switches to detect when a plug is inserted into the jack, bringing up a software configuration menu.

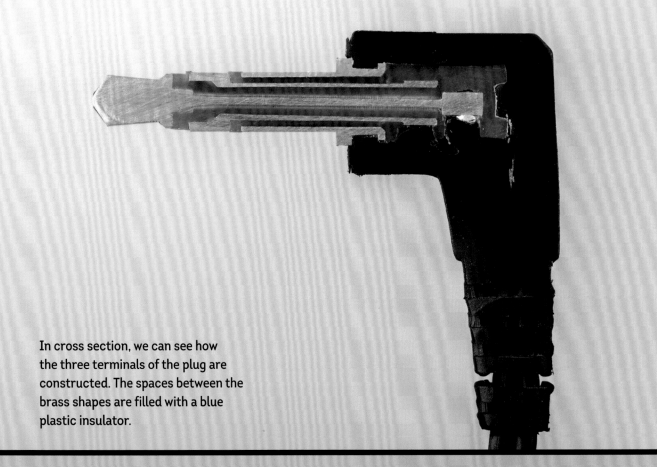

In cross section, we can see how the three terminals of the plug are constructed. The spaces between the brass shapes are filled with a blue plastic insulator.

Spring-loaded contacts in the jack disconnect two switches, retain the plug in place, and make consistent contact to the tip, ring, and sleeve.

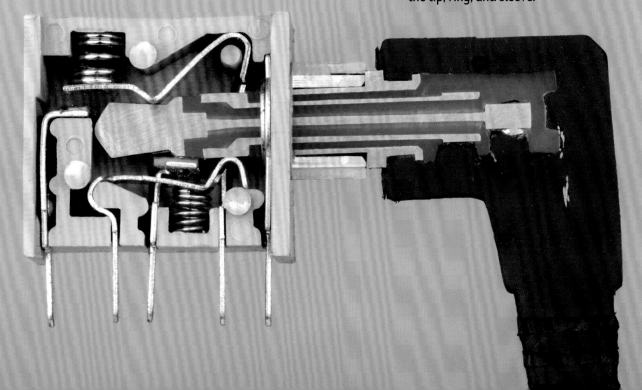

LMR-195 Coaxial Cable

A **COAXIAL CABLE** has two conductors inside: a central wire that carries the signal, and an outer, braided shield that carries ground currents and protects the signal from interference. The term *coaxial* ("co-axial"), or sometimes just *coax*, signifies that both conductors share the same center axis.

The cable is designed to carry radio frequency (RF) signals. It's easy to think of the RF as traveling down the center conductor, but it actually moves along the gap between the center conductor and the outer shield.

The shield is more than a loose braid of copper wires: there are multiple layers of wires laid across one another, and there's an additional aluminum foil wrap between those wires and the plastic insulation around the center conductor. These features all help improve the characteristics of this high-quality LMR-195 cable, named for being 0.195 inches in diameter.

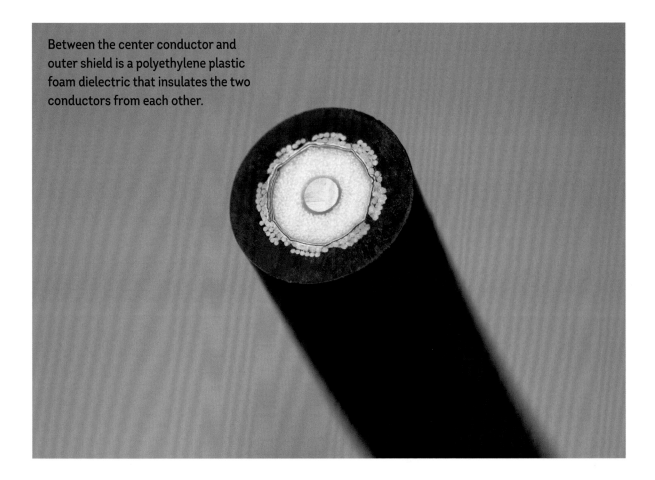

Between the center conductor and outer shield is a polyethylene plastic foam dielectric that insulates the two conductors from each other.

Laptop Power Cable

This is the power cable from an Apple MacBook Pro computer. From the outside, the cable is a soft white rubber noodle, flexible and easy to grip. The flexibility comes from the immense number of thin and supple copper strands that can flex and slide a little bit against one another.

The straight inner bundle of wires is packed around a **STRENGTH MEMBER**, a cord of high-strength fiber, possibly Kevlar. A firm plastic insulator around the inner bundle keeps the cable from being bent too sharply. The outer wire bundle is the ground return for power transmission. It consists of opposing spirals of copper strands, which can more easily adapt to changes in length than the straight strands in the middle, assisting with flexibility. The outer rubber jacket is a tough and textured polymer with a rubbery finish.

This cable is designed to be flexible yet tough and strong enough to survive its role as an occasional trip hazard.

RG-6 Coaxial Cable

Between a cable modem and its wall outlet runs a coaxial cable such as this one.

While the general design of the cable is similar to the previous LMR-195 coaxial cable, this RG-6 cable is designed to be lower cost and differs in the details of its construction. For example, it uses aluminum wires for shielding instead of copper and the center conductor is plated with copper instead of being made from solid copper.

The shielding on this cable has two layers of aluminum foil around an aluminum-wire braid.

Cable TV RG-59

The price difference between a mid-priced cable and the cheapest ones available can be substantial, but so too can be the quality. You often can't spot any problems from the outside; to really see the difference, you have to cut a cable in half, like we've done here.

In comparison to high-quality coaxial cables, this cable TV cable isn't just inexpensive but also poorly manufactured. The center conductor isn't in the middle of the plastic dielectric, the outer shield is just a few sparse strands of wire, and the plastic jacket thickness is quite inconsistent. The inconsistent profile

and poor shielding both suggest that this cable will underperform compared to its peers.

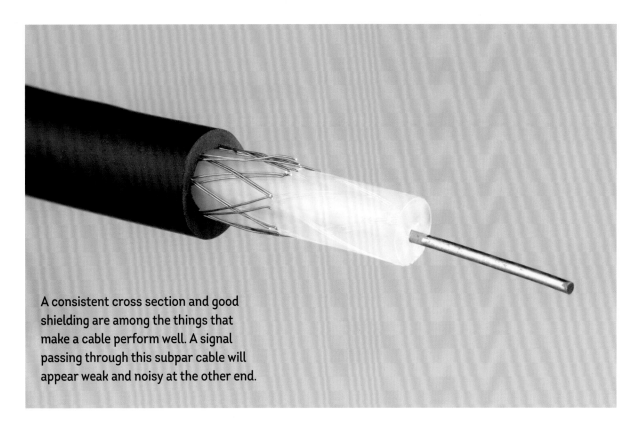

A consistent cross section and good shielding are among the things that make a cable perform well. A signal passing through this subpar cable will appear weak and noisy at the other end.

F Connector

An **F CONNECTOR** is the threaded connector you can find on the back of a TV box or cable modem. A somewhat unusual feature of this connector is that its center "pin" is actually just the protruding center conductor of the coaxial cable itself.

When the cable is plugged into the jack, the center conductor is pinched in a spring-like contact that makes the electrical connection and brings the signal into (or out of) the electronic device. A plastic spacer helps guide the center conductor into the contact.

The outer hex nut is free to rotate and fastens the plug onto the connector. Some F connectors use a "push on" design with springs that grip the threads so that the connector can simply be pushed into place.

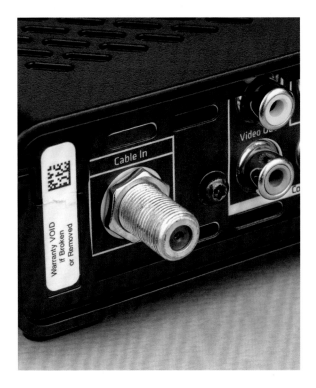

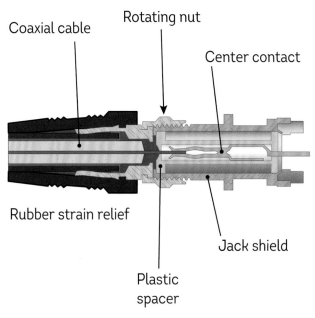

Coaxial cable

Rotating nut

Center contact

Rubber strain relief

Jack shield

Plastic spacer

An F connector plug and jack are typical connectors used for cable TV and internet.

The solid center conductor of the coaxial cable serves as the center pin of the connector plug.

BNC Plug and Jack

The **BNC CONNECTOR** is a very popular coaxial cable connector for radio frequency signals and general laboratory use. Unlike the F connector, it only takes a quick quarter-twist of its bayonet-style mount to connect and disconnect.

BNC connectors, like most coaxial connectors, use a crimped center pin, rather than the bare wire itself, to make the central connection. And, like other high-quality coaxial connectors, the BNC is designed to present a relatively constant impedance along its body.

Loosely speaking, **IMPEDANCE** refers to the amount of effective resistance that a circuit presents to both DC and AC signals. Cables and connectors with constant impedance minimize unwanted echoes in transmitted signals.

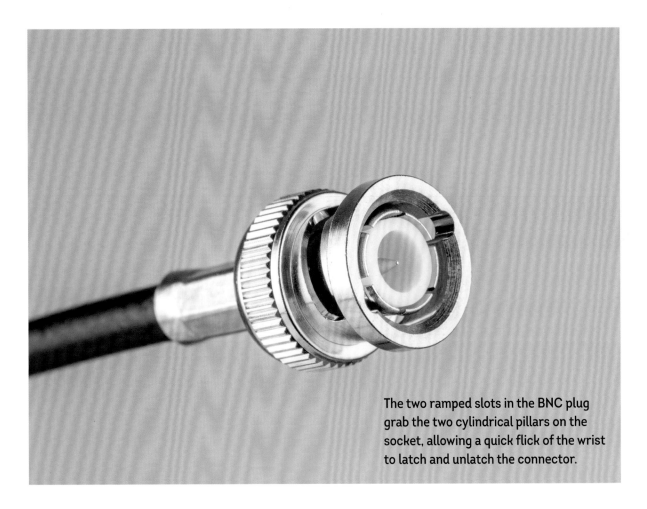

The two ramped slots in the BNC plug grab the two cylindrical pillars on the socket, allowing a quick flick of the wrist to latch and unlatch the connector.

BNC connectors are typically machined in nickel-plated brass.

This plug and jack share a fairly consistent coaxial profile of the center conductor and dielectric in the section where they meet.

SMA Connector

Fancy high-tech gear like signal generators use small, precise SMA CONNECTORS. They transmit their signal much more faithfully than consumer-grade connectors.

The cable shown with the connector is called "semi-rigid" because its outside shield is a hollow tube of tin-plated copper. Semi-rigid coax isn't flexible, but can be bent into the necessary shape using specialized tools. Other than the solid shield, it's a regular coaxial cable with a center conductor and dielectric plastic around it.

An SMA plug has a freely rotating hexagonal nut that can be tightened onto the matching jack.

In cross section, we can see quite a bit of detail thanks to the different types of metal used. The outer SMA plug parts are stainless steel, while the jack is made from gold-plated brass.

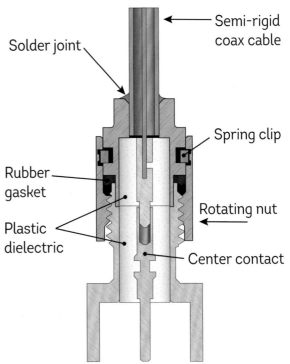

Solder joint

Semi-rigid coax cable

Spring clip

Rubber gasket

Rotating nut

Plastic dielectric

Center contact

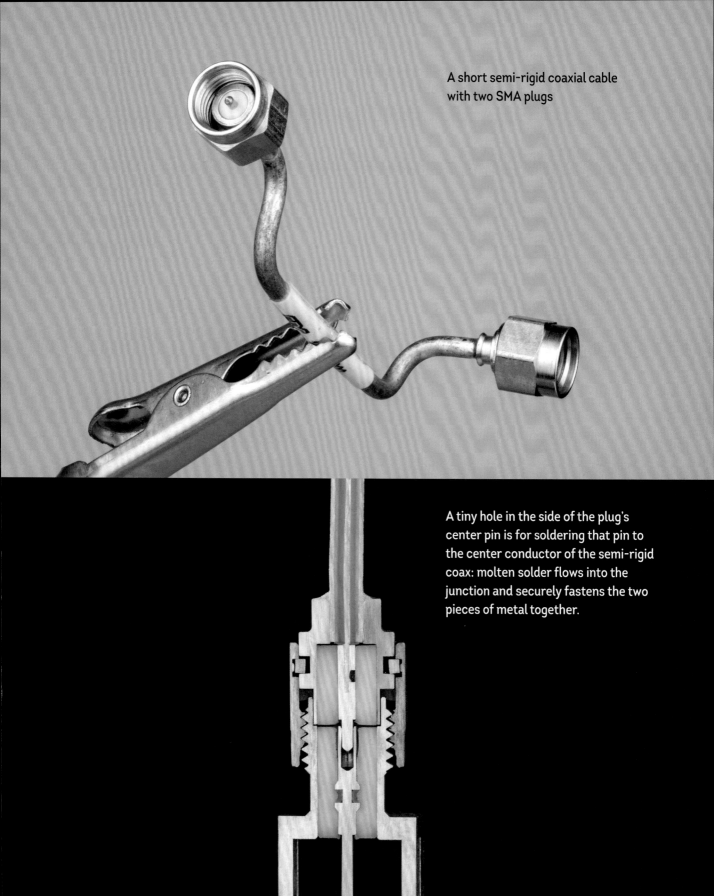

A short semi-rigid coaxial cable
with two SMA plugs

A tiny hole in the side of the plug's
center pin is for soldering that pin to
the center conductor of the semi-rigid
coax: molten solder flows into the
junction and securely fastens the two
pieces of metal together.

DE-9 Connector

Older computers used a **DE-9 CON-NECTOR** for carrying serial data in the RS-232 protocol. Since a lot of computers and instruments still use this old data transmission standard, DE-9 cables and adapters can still be purchased.

These simple, robust connectors have nine pins on the plug that fit neatly into nine spring-loaded sockets on the receptacle. The trapezoidal metal shells guide the connectors into alignment,

preventing damage to the pins when connecting them.

The DE-9 connector is frequently yet incorrectly called a "DB-9" connector. This is likely because of its similarity to the wider DB-25 connector that was used for parallel port printers and older serial connections. The "B" or "E" refers to the size of the connector shell; DE-9 is the correct name for this smaller connector.

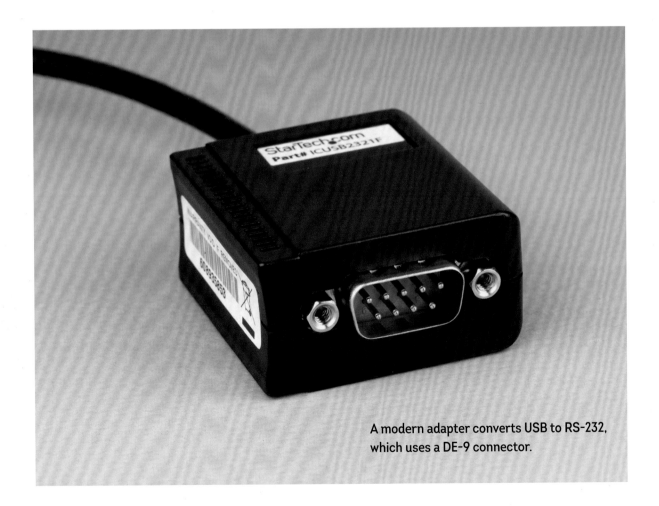

A modern adapter converts USB to RS-232, which uses a DE-9 connector.

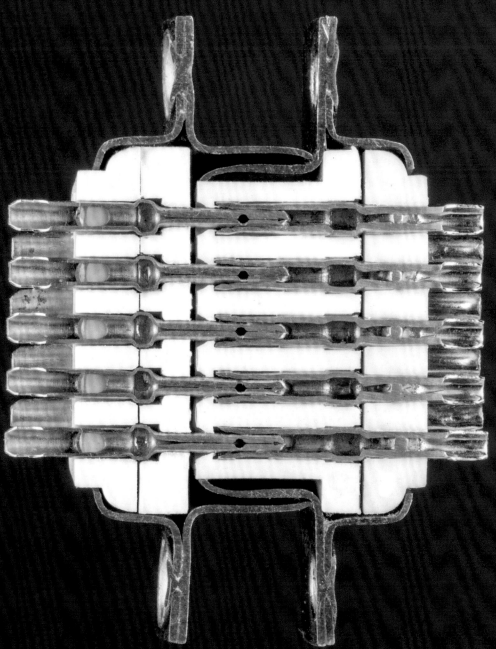

A DE-9 plug and receptacle in cross section. In practice, wires would be soldered to each of the nine terminals on each side.

Category 6 Ethernet Cable

The **CATEGORY 6 (CAT6) ETHERNET CABLE** contains four twisted pairs of copper wire. These common cables are used worldwide to carry local network and internet traffic.

Cat6 cables improve on previous generations of cables by adding an internal X-shaped plastic spacer to keep the pairs of wires apart from each other, reducing the signal leakage between pairs. They also have a metal foil shield around the outside to reduce interference from external signals, as well as a separate "drain" conductor to assist with shielding.

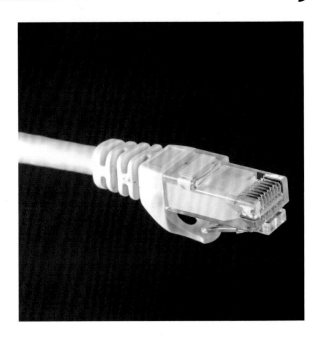

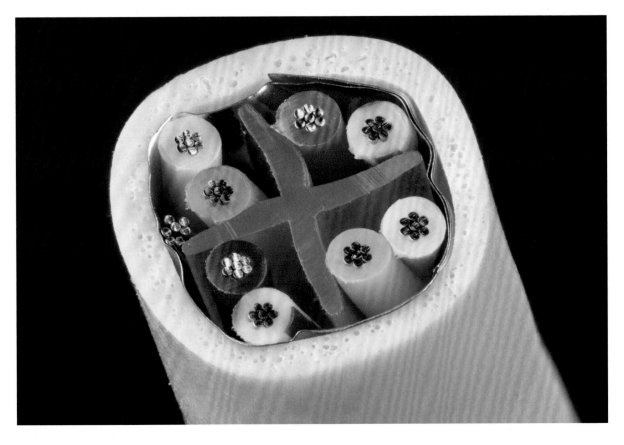

SATA Cable

SATA (SERIAL AT ATTACHMENT) CABLES are used inside computers to connect internal storage drives to the motherboard. Each cable contains two **TWINAXIAL** pairs. One pair carries data to the hard drive, and the other pair carries data from the hard drive.

A twinaxial, or twinax, cable looks like two coaxial cables stuck together, sharing the same outer shield. Signals are transmitted with **DIFFERENTIAL SIGNALING**, meaning signals are represented by the voltage difference between the two wires. This system effectively cancels out most electrical interference, since the interference adds equally to both signal wires.

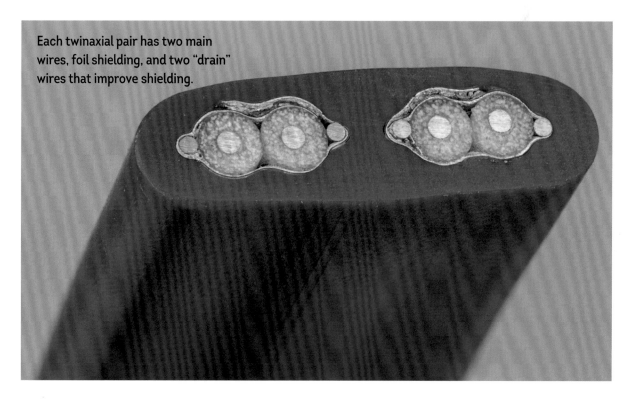

Each twinaxial pair has two main wires, foil shielding, and two "drain" wires that improve shielding.

HDMI Cable

HIGH-DEFINITION MULTIMEDIA INTERFACE (HDMI) CABLES connect computers and other video devices to monitors and televisions.

The cable has four independently shielded twisted pairs of wires that transmit data, including digital video. The stream of video data gets split into four separate streams of serial digital data, one per pair of wires. At the video monitor end, the four streams are combined and decoded to generate the picture.

Other unshielded wires carry low-speed auxiliary signals that identify the make, model, and resolution of the display, or allow remote control of volume and other settings. The overall cable is shielded with layers of aluminum foil and a copper braid.

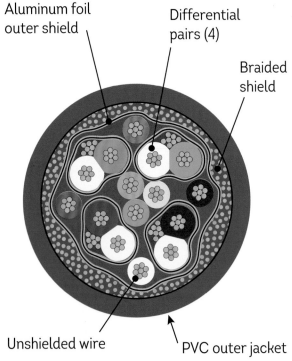

Aluminum foil outer shield

Differential pairs (4)

Braided shield

Unshielded wire

PVC outer jacket

Each signal pair is shielded with a foil wrap and copper drain wire.

VGA Cable

Before HDMI and DisplayPort, in the old days of analog video, **VIDEO GRAPHICS ARRAY (VGA) CABLES** were used to carry video signals from computers to monitors.

VGA uses analog video signals consisting of three electrical signals, representing the color components red, green, and blue. For the two VGA cables shown here, the manufacturers have helpfully color-coded the three miniature coaxial

cables that transmit these signals. One of the cables also contains a smaller gray coax for transmitting horizontal synchronization information, but the other VGA cable uses a regular wire for the same purpose.

VGA cables also have other wires, used for other sync signals and to transmit auxiliary information, such as data to identify the monitor.

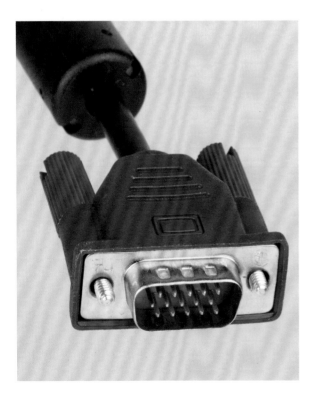

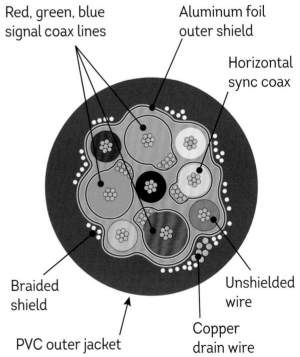

Red, green, blue signal coax lines

Aluminum foil outer shield

Horizontal sync coax

Braided shield

PVC outer jacket

Unshielded wire

Copper drain wire

The outer shield of this cable is aluminum, with a stranded copper drain wire added.

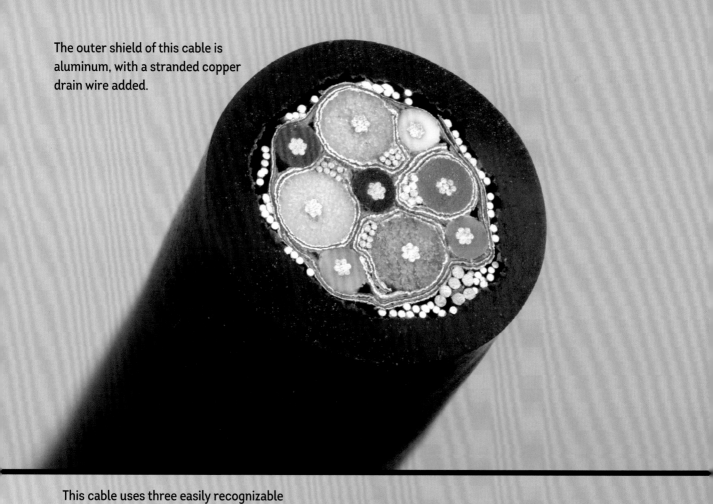

This cable uses three easily recognizable miniature coaxial cables for its red, green, and blue components.

Basic USB Cable

If you've used a computer recently or even just charged your phone, you've probably used a **USB CABLE**.

Cutting into a USB cable and connector reveals a number of interesting details. The plug uses insulation-displacing contacts, similar to those that we saw in the ribbon cable. A pair of tiny metal teeth pops through the insulation of each wire and makes electrical contact with the conductive copper inside.

Within the cable, we can identify two larger red and black wires that carry power, as well as two smaller white and green signal wires. All are seven-strand wires: the larger wires are made with larger strands, rather than additional strands. Layers of wrapped foil and aluminum braided wire shield the signals, preventing interference. The outer protective jacket is made of PVC plastic.

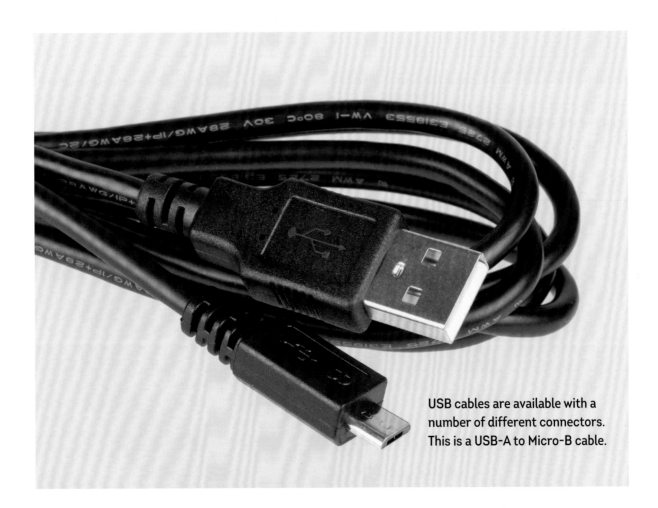

USB cables are available with a number of different connectors. This is a USB-A to Micro-B cable.

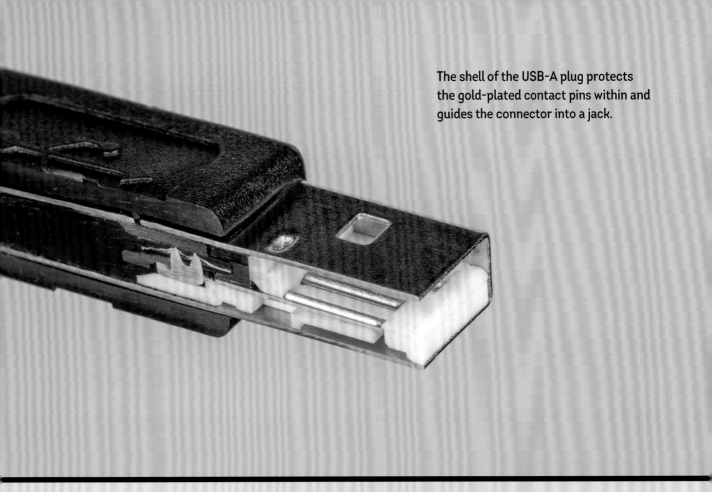

The shell of the USB-A plug protects the gold-plated contact pins within and guides the connector into a jack.

A basic USB cable contains only a few wires for signal and power. Newer "SuperSpeed" USB cables are much more complex.

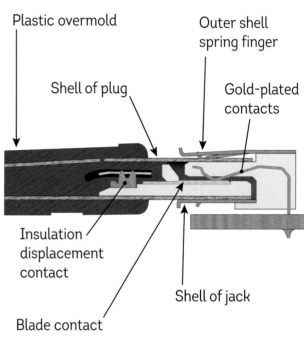

Plastic overmold

Outer shell spring finger

Shell of plug

Gold-plated contacts

Insulation displacement contact

Shell of jack

Blade contact

USB Jack

The gold-plated contact fingers in the USB jack act like springs, pressing down tightly against the flat metal blade contacts inside the plug.

Spring fingers on the outer shell lock into recesses on the plug, helping to prevent the cable from falling out after it's been plugged in. The "clicking" sensation you feel when you plug your cable in comes from these spring fingers.

If it ever feels like it takes several tries to plug in a USB cable, that's because of these spring fingers. (You had it right the first time; they were just a bit too stiff.)

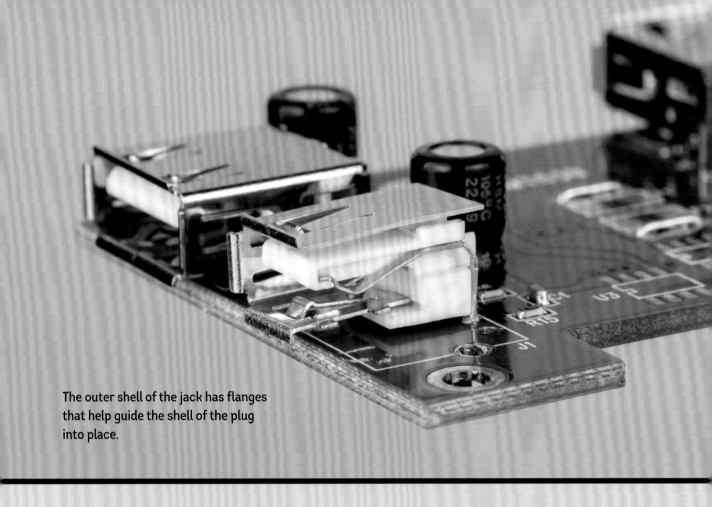

The outer shell of the jack has flanges that help guide the shell of the plug into place.

The gold-plated spring fingers in the jack press firmly against the matching contacts in the plug.

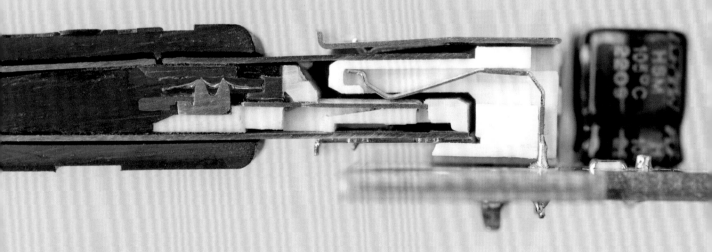

SuperSpeed USB Cable

A top-of-the-line 10 Gbps SuperSpeed USB cable is a tiny, precisely constructed, dazzling work of art.

Most remarkable are the eight miniature shielded coaxial cables, each just 1 mm in diameter, with its own color-coded foil wrap. Each pair of two coax cables forms a high-speed data transmission lane. With four lanes total, the USB cable can move data up to 10 gigabits per second.

Near the center, the cable has thick red and black wires for powering devices, and a shielded green and white signal pair. In a sense, these form a low-end basic USB cable embedded within the high-end SuperSpeed cable for backward compatibility.

Four smaller wires near the outer shield carry auxiliary signals. The whole cable is wrapped with an outer braided copper shield for improved immunity to electrical interference.

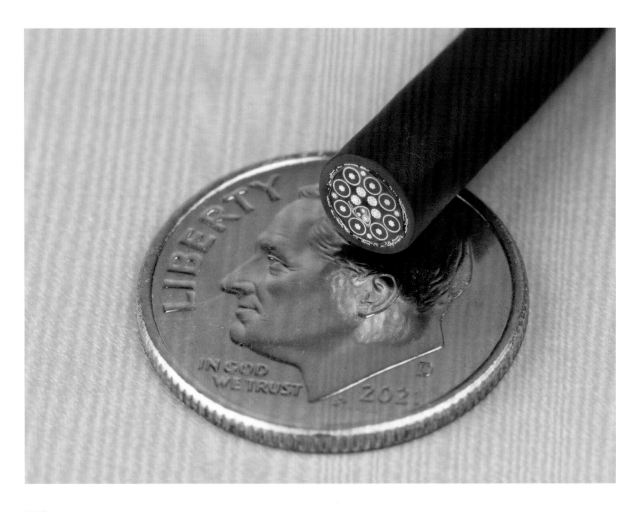

In addition to electrical connections, this cable has a strength member, a strong fiber like Kevlar, visible as the yellow-colored area near the center, between the red and black power conductors.

5

Retro Tech

Some of the most iconic electronic components are, simply put, obsolete. Photo flashbulbs have given way to LEDs, Nixie tubes have (in every sense but aesthetic) been replaced by seven-segment displays, and analog panel meters have been supplanted by digital displays. Some of the retro items we'll examine, like core memory, have been out of use for decades, while others, like incandescent light bulbs, are on the cusp of obsolescence. A rare holdout is the vacuum tube, still commonly manufactured for use in guitar amplifiers.

Neon Lamp

NEON LAMPS contain a small amount of the eponymous noble gas. When enough voltage is applied between the lamp's two parallel electrodes, the gas ionizes and emits a distinctive orange glow.

The outer glass envelope holds the electrodes in place and prevents the neon gas from escaping. The blob of glass at the tip is where the envelope was sealed after introduction of the gas during the manufacturing process.

When a DC voltage is applied across the leads, only the negative electrode (cathode) lights up. With an AC voltage, which alternates between positive and negative polarities many times per second, each electrode lights up in turn. Due to persistence of vision, both electrodes appear lit.

Neon lamps are often used as indicators for AC power—for example, in extension cords, light switches, and power switches.

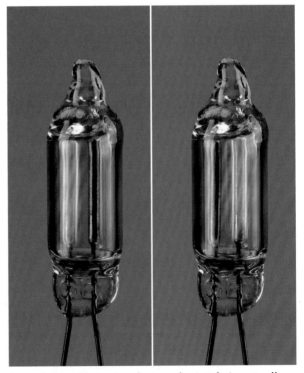

At any given instant, only one electrode is actually lit. Because they alternate quickly, it looks as though both are lit.

This neon lamp is about 6 mm in diameter. Its base is stamped GE, for General Electric.

Nixie Tube

Before seven-segment LED displays were common, manufacturers used distinctive, neon-filled **NIXIE TUBES** to display numeric information.

Like a neon lamp, the gas inside a Nixie tube ionizes when the electrodes are connected to a high voltage. Unlike a neon lamp, each negative electrode has been formed into the shape of a numeral. The glow around each illuminated digit is broad enough that the electrodes can be stacked into a compact array without occluding each other. The Nixie tube shown here has two anodes: the hexagonal grid in front of the digits and the solid metal backshell behind them.

Although the original manufacturers of Nixie tubes shut down their factories decades ago, so many people love the friendly orange glow of this unique display technology that new factories have begun building Nixie tubes again.

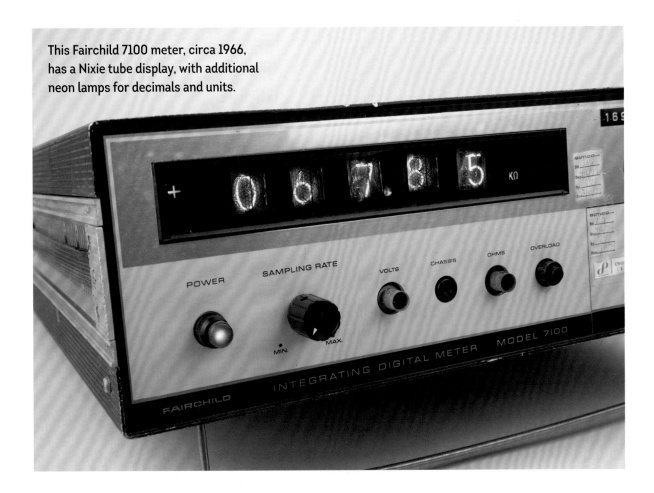

This Fairchild 7100 meter, circa 1966, has a Nixie tube display, with additional neon lamps for decimals and units.

A ZM1030 Nixie tube has an orange coating to increase the display contrast.

Inside the Nixie Tube

With the Nixie tube's glass envelope and hexagonal front anode grid removed, we can see the shaped cathodes inside. The digits are stacked and separated by insulating ceramic washers.

This Nixie tube has fewer pins than digits. The odd numerals are in the front half, and are lit via the front anode grid. The even numerals are in the back half, and are lit via the rear anode, the black metal

shell behind the digits. The cathodes are wired together in pairs such that (for example) the 0 and 1 digits are connected together, but only one of them lights at a time depending on which anode is active.

A transparent screen of almost impossibly fine tungsten wire, roughly 0.01 mm thick, spans the tube between the front and rear halves to keep the influence of each anode limited to its half.

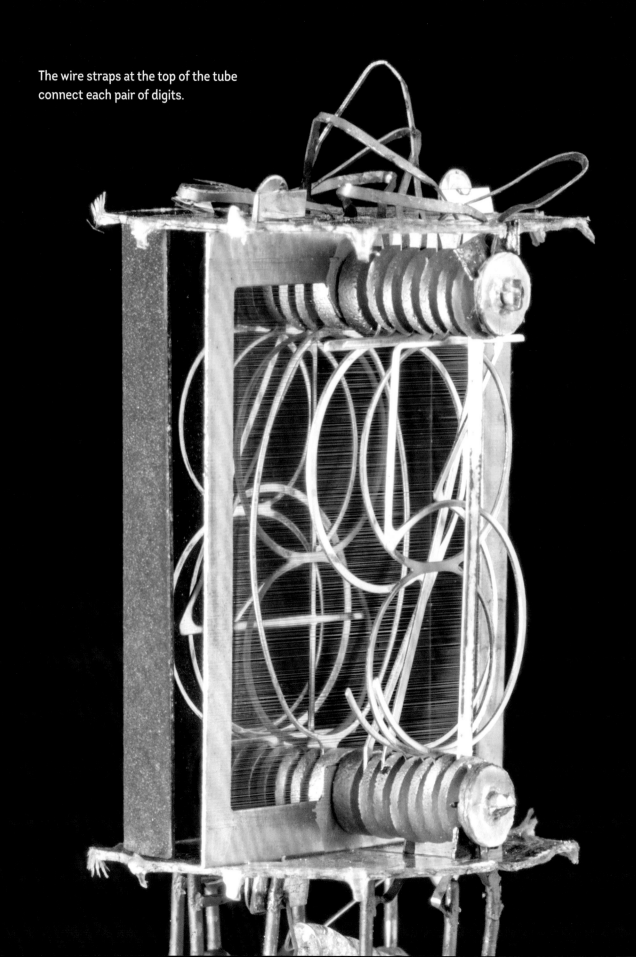

The wire straps at the top of the tube
connect each pair of digits.

12AX7 Vacuum Tube

Known to audio enthusiasts and guitar players the world over, the iconic **12AX7 VACUUM TUBE** has been amplifying signals since the 1940s.

In cross section, you can immediately see that there are two copies of the same thing inside; the 12AX7 is a **DUAL TRIODE** vacuum tube that can amplify two signals at once. Each of the two **TRIODES** has three elements: the inner cylindrical **CATHODE**, the wire **GRID**, and the outer **PLATE**. Underneath is a mica washer that insulates and supports the elements.

When operating, the cathodes are heated by tiny resistive filaments, emitting the characteristic warm glow of vacuum tubes. Electrons emitted by the cathode flow toward the plate, but can be repelled by a small voltage applied to the grid. Thus, small signals applied to the grid can be amplified into a larger output at the plate.

The vacuum in the tube allows the electrons to flow freely, without interacting with air molecules.

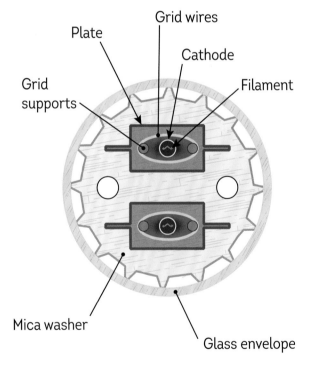

Plate

Grid wires

Cathode

Filament

Grid supports

Mica washer

Glass envelope

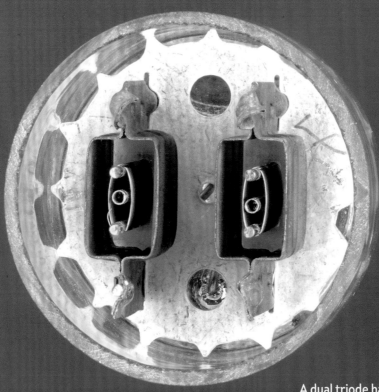

A dual triode has two copies of the main structure within.

The upper part of this vacuum tube, including a second mica support washer, was removed for photography.

Vacuum Fluorescent Display Tube

A **VACUUM FLUORESCENT DISPLAY**, or **VFD**, is a special type of vacuum tube for displaying information. Despite the rise of seven-segment displays, VFDs are still widely used in car dashboards and household appliances.

Thin, wide vacuum tubes with flat glass faces, VFDs are low-voltage devices, essentially triode vacuum tubes—like 12AX7s—but their anode plates are coated with a phosphor. The cathodes are a set of six very fine heated filament wires strung tightly across the front of the display. Beneath the cathodes are control grids etched from very thin sheets of metal. Below those are the phosphor-coated anode plates that form the visible display elements.

The filaments release electrons. When voltage is applied to a particular phosphor-coated anode, it attracts electrons and emits the familiar fluorescent-green glow.

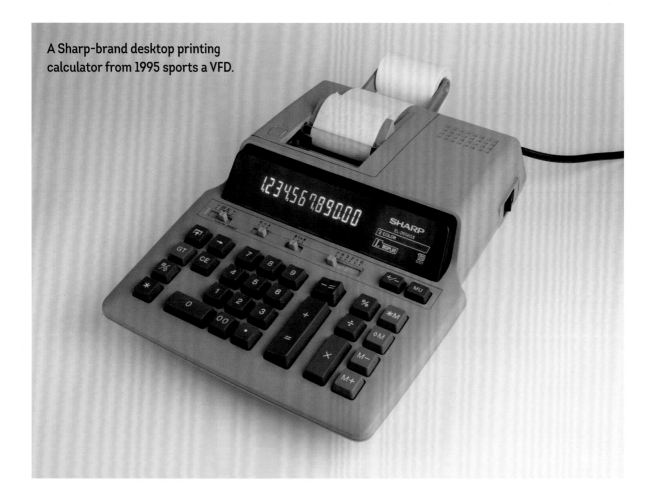

A Sharp-brand desktop printing calculator from 1995 sports a VFD.

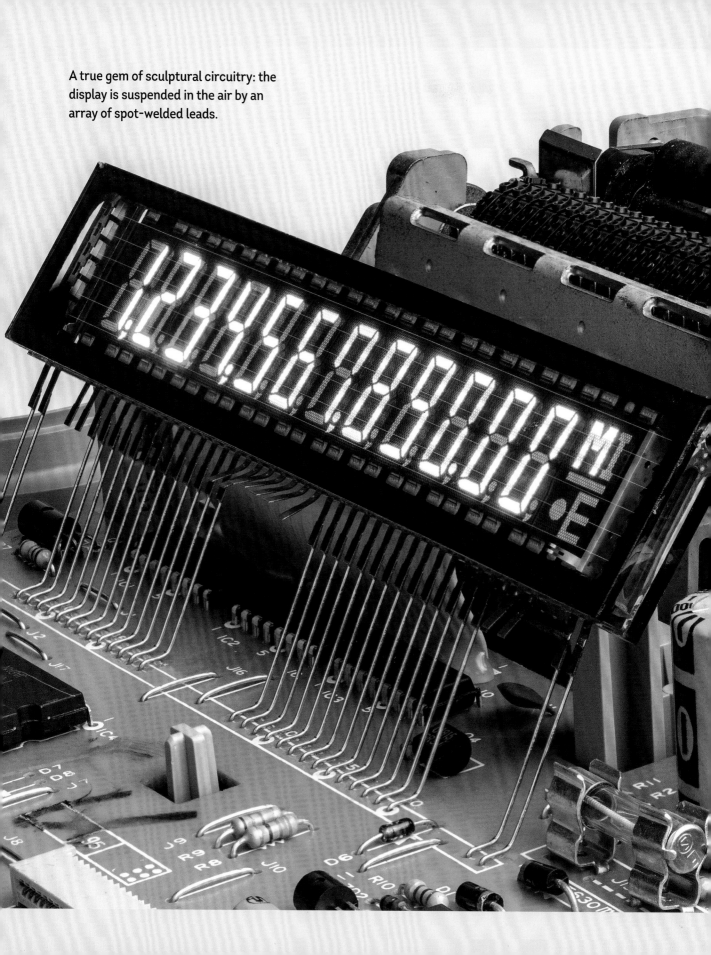

A true gem of sculptural circuitry: the display is suspended in the air by an array of spot-welded leads.

Cathode Ray Tube

A **CATHODE RAY TUBE**, or **CRT**, was once the display screen for every television set and computer monitor. The name is a historical artifact referring to "rays" emitted by a heated cathode; we now call those rays *electrons*.

CRTs generate images through the same process we just saw in the vacuum fluorescent display: electrons hitting a phosphor in a vacuum. Specifically, a CRT is a vacuum tube where an **ELECTRON GUN** generates an extremely fine beam of electrons that stream toward a phosphor-coated screen, which glows wherever the beam hits it. Electromagnets around the tube steer the beam around the screen to build up a complete image one line at a time, like someone methodically mowing a lawn.

The black-and-white CRT that we examine here is quite small, of the type used for video camera viewfinders.

This CRT viewfinder is from a 1990s JVC brand camcorder.

The round CRT inside the viewfinder is surrounded by a magnetic yoke. A connector brings electrical signals to the pins of the tube.

The magnetic yoke contains adjustable ferrite pieces as well as perpendicular deflection coils for steering the beam up, down, left, and right.

The coiled filament is stretched across and just to the right of the gap between the two wires at the center of this photo.

Inside the CRT

At the core of the CRT is an electron gun, a specialized vacuum tube component that produces a focused beam of electrons.

It all begins with a heated filament. For this tiny CRT, the filament is made of ultra-fine wire about 0.01 mm in diameter. The wire is coiled and stretched across a gap of 0.7 mm, about the

thickness of seven sheets of paper. The heated filament throws off electrons, which are then focused and accelerated toward the screen using high voltages applied to a series of cup-shaped electrodes. As the electrons leave the electron gun, they're deflected by the magnetic yoke to the correct positions on the phosphor screen.

The front of the tube is a flat surface, coated on the inside with a phosphor.

In cross section, we can see all the parts of the CRT, from the electron gun on the left to the phosphor screen on the right.

Mercury Tilt Switch

In this simplest of switches, a tiny blob of conductive, liquid mercury metal can complete an electrical circuit by making contact with the two electrodes, but only when the device is upright. When tilted downward, the mercury falls away from the electrodes, opening the circuit.

These switches are no longer readily available due to mercury's toxicity, but years ago you could find them in simple electromechanical thermostats. The mercury switch was secured to the end of a coiled bimetallic strip. Heating or cooling the strip rotated the mercury switch until, at a set temperature, the switch closed, and the heater or air conditioner turned on.

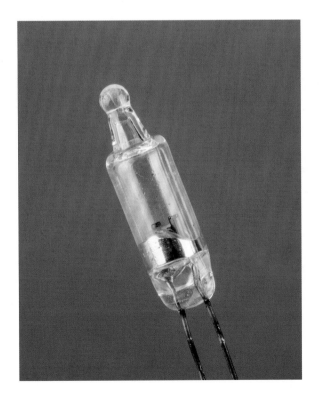

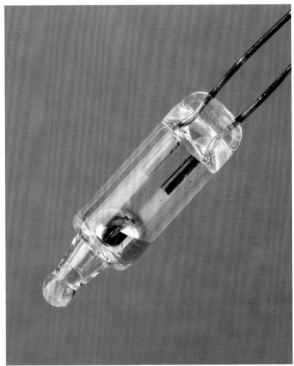

Vintage Wirewound Resistors

Small carbon resistors can't handle much power; beyond a few watts, they overheat, and the fragile carbon coating disintegrates. As we saw on page 19, larger power resistors are made from a wound resistive wire and encapsulated in ceramic packages that can handle high temperatures. Both of the wirewound resistors shown here are classic designs dating back nearly a century, though versions of each are still being manufactured today.

The **TUBULAR VITREOUS ENAMEL RESISTOR** is a classic style of robust, low-cost wirewound power resistor. Ones like this, with no enamel on the side, can be used with a clamped-on wiper to act as a crude potentiometer.

The **MICA CARD RESISTOR** is used for highly stable circuits that need to work at a wide range of power dissipation values. It has a length of thin resistive wire that can be precisely calibrated, looped many times over a heat-tolerant mica form.

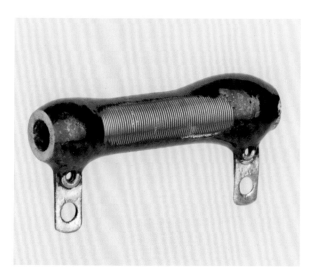

Tubular vitreous enamel resistor

Mica card resistor

Carbon Composition Resistor

CARBON COMPOSITION RESISTORS are often found in older electronics, such as antique radio equipment. The resistive element is a carbon "composition," what we more commonly refer to as a **COMPOSITE**. It starts as a thick paste made from conductive carbon powder, non-conductive ceramic clay, and a binder resin.

After curing, the composite has the appearance of a terrazzo floor. Pale grains of clay stand out against the dark carbon-bearing resin. The outer shell is molded from a **PHENOLIC** resin such as Bakelite.

Notice how the composite grain is distorted near the ends of the connecting wires. The wires were pushed into the paste while it was soft, prior to curing.

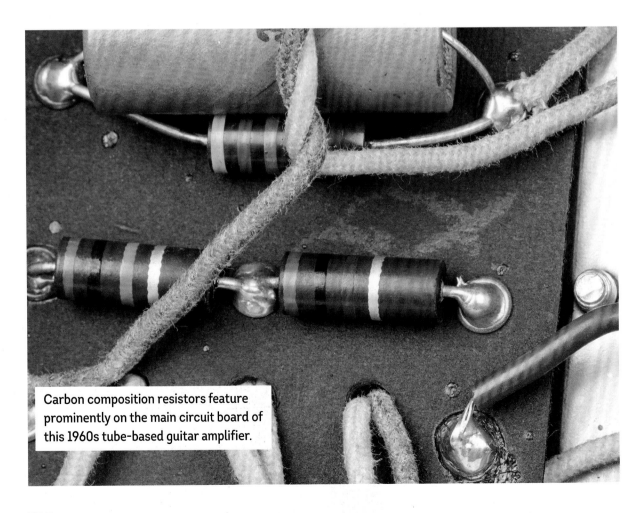

Carbon composition resistors feature prominently on the main circuit board of this 1960s tube-based guitar amplifier.

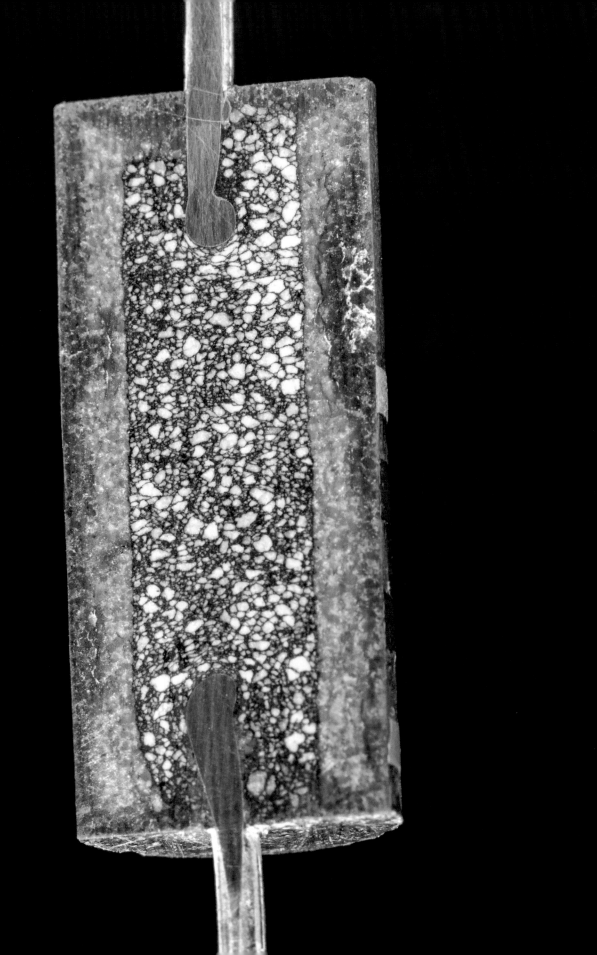

Cornell-Dubilier 9LS Capacitor

The dielectric material in this 1920s capacitor is mica, a naturally occurring mineral that can be as transparent as glass. Mica is an electrical insulator that can readily be cleaved into parallel sheets of uniform thinness.

Soft metal plates form the capacitor's electrodes. To increase the capacitance, multiple layers of alternating metal plates and mica sheets are stacked together. This "sandwich" is compressed tightly and impregnated with an insulating compound before being attached to the threaded inserts that act as contact points. Finally, the whole assembly is molded in Bakelite to protect the fragile internal components.

W. DUBILIER.
CONDENSER AND METHOD OF MAKING THE SAME.
APPLICATION FILED OCT. 30, 1918.

1,345,754.   Patented July 6, 1920.

Fig.1.

Inventor
William Dubilier

The metal wiring from each end of the capacitor is wrapped around the screw terminals to make an electrical connection.

With a foreshortened perspective, we can see the interdigitated layers of this capacitor: shiny silver layers separated by darker mica layers.

Dipped Silver Mica Capacitor

Instead of using separate soft metal plates and mica sheets, this type of **SILVER MICA CAPACITOR**, invented around the 1950s, uses a special plating process to deposit silver directly on the surface of the mica insulator. As with the Cornell-Dubilier capacitor, multiple plated sheets are stacked together for more capacitance.

Thin layers of metal foil between the mica sheets connect the plated silver electrodes to the two large, brass-colored metal clamps crimped onto the stack. The completed capacitor is encapsulated in a phenolic resin to protect it.

Axial Multilayer Ceramic Capacitor

On page 36, we examined a multilayer ceramic capacitor that mounts directly onto a printed circuit board. For a time, MLCCs were also available sealed inside tiny glass tubes with attached wires so they could be installed by soldering the wires into the plated through holes of a circuit board, much like an ordinary axial resistor or diode.

The glass envelope, connections, and sealing techniques used here are similar to those used for the glass-encapsulated diodes on page 66.

This axial-packaged MLCC, with a passing resemblance to a strawberry ice cream sandwich, has about 30 interdigitated metal layers.

IF Transformer

An **IF TRANSFORMER**, short for *intermediate frequency*, is a tunable inductor that often has a built-in capacitor. These were once very popular in television sets and radios, like the 1960s transistor radio circuit board in the photo.

The inductance of an inductor depends on the type and position of its magnetic core, and the IF transformer's core is a movable, screw-shaped ferrite slug. As the slug is rotated—with a special plastic tool that won't crack the brittle ferrite—it moves up or down within the inductor windings. This changes the characteristics of the IF transformer and adjusts the response of the circuit that uses it. Having a built-in capacitor allows the circuit designer to save an external component.

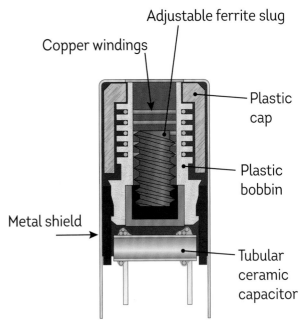

Copper windings

Adjustable ferrite slug

Plastic cap

Plastic bobbin

Metal shield

Tubular ceramic capacitor

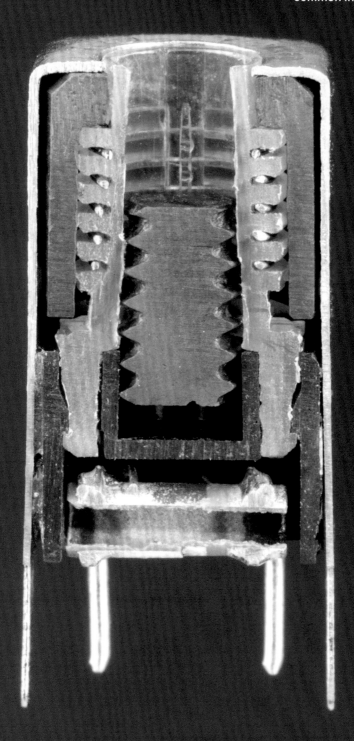

The cut silvery tube at the bottom of the IF transformer is an old-style tubular ceramic capacitor, a type common in the 1960s.

Incandescent Light Bulb

The **INCANDESCENT LIGHT BULB** is the classic light bulb, where a hot tungsten filament glows simply because it is hot. These bulbs aren't particularly efficient as light sources because so much of their power output is released as heat rather than light, which is why they're now being replaced by LEDs.

The filament of the bulb, at first glance or even under moderate magnification, appears to be a simple coil of wire. Under high magnification, however, it turns out to be a coiled coil of much finer wire.

The ends of the filament are cool enough that they don't glow, in part because the two vertical supports act as heat sinks, wicking heat away from the ends of the filament.

Switching on a light bulb generates a thermal pulse, which puts a lot of stress on the filament. This is why light bulbs preferentially burn out right as you throw the switch.

The first readily available light bulbs used a fine wisp of carbon as the filament. These were soon replaced by tungsten filaments, which last longer and are cheaper to manufacture.

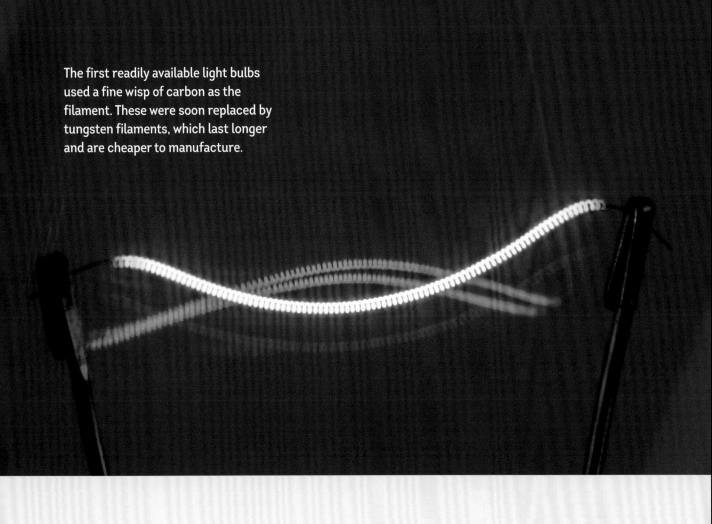

The filament is a coiled coil of remarkably fine tungsten wire.

Camera Flashbulb

While incandescent light bulbs are designed to last as long as possible before burning out, single-use **FLASH-BULBS** are designed to burn out the *very instant* they're first turned on—just enough time to expose photographic film for that glamor shot.

Instead of a tungsten filament, flash-bulbs contain ribbons, wires, or foils of magnesium, packed in an oxygen-filled glass envelope. When a pulse of voltage is applied, the magnesium heats up and ignites, burning very rapidly in the oxygen-rich atmosphere and emitting a bright flash of light that lasts a fair bit longer than our modern strobe-based camera flashes.

The outer blue plastic coating on these bulbs filters the output light and strengthens the glass envelope.

These Press 25B-size flash bulbs are about 25 mm in diameter and are packaged a dozen per wonderful box.

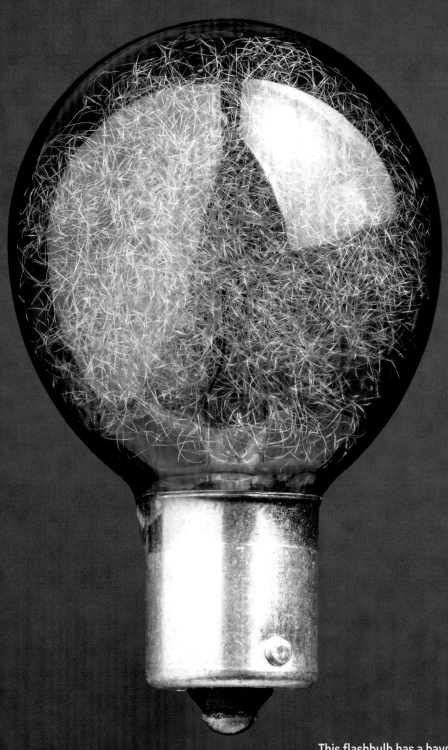

This flashbulb has a bayonet-style base
for quick changes in a camera flash unit.

Photoresistor

A **PHOTORESISTOR** is known by a number of other names: **PHOTOCELL**, **CdS CELL**, or **LIGHT-DEPENDENT RESISTOR (LDR)**. It's a circuit element that acts like a resistor, but the resistance changes depending on the amount of light that hits it.

Photoresistors are made from a ceramic substrate coated with cadmium sulfide (CdS) or cadmium selenide (CdSe). Metal electrodes are placed on top in a distinctive interdigitated pattern. The central wiggly line is the long, narrow gap between the two electrodes, which exposes the cadmium compound below.

The cadmium compound may be bright yellow or red; CdS and CdSe are also known as the artists' pigments **CADMIUM YELLOW** and **CADMIUM RED**. Both compounds change their resistivity with the presence of light.

Cadmium compounds are toxic, and photoresistors like these are gradually being phased out in favor of silicon light sensors.

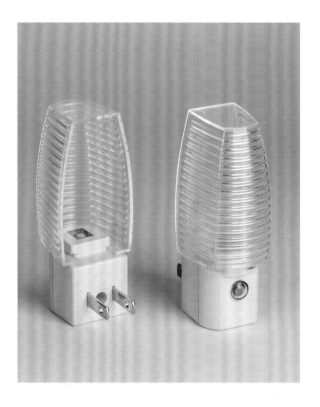

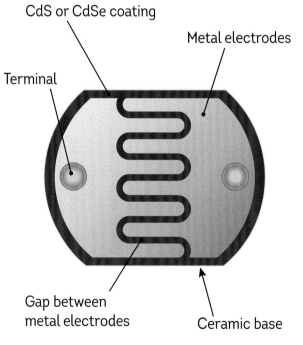

CdS or CdSe coating

Metal electrodes

Terminal

Gap between metal electrodes

Ceramic base

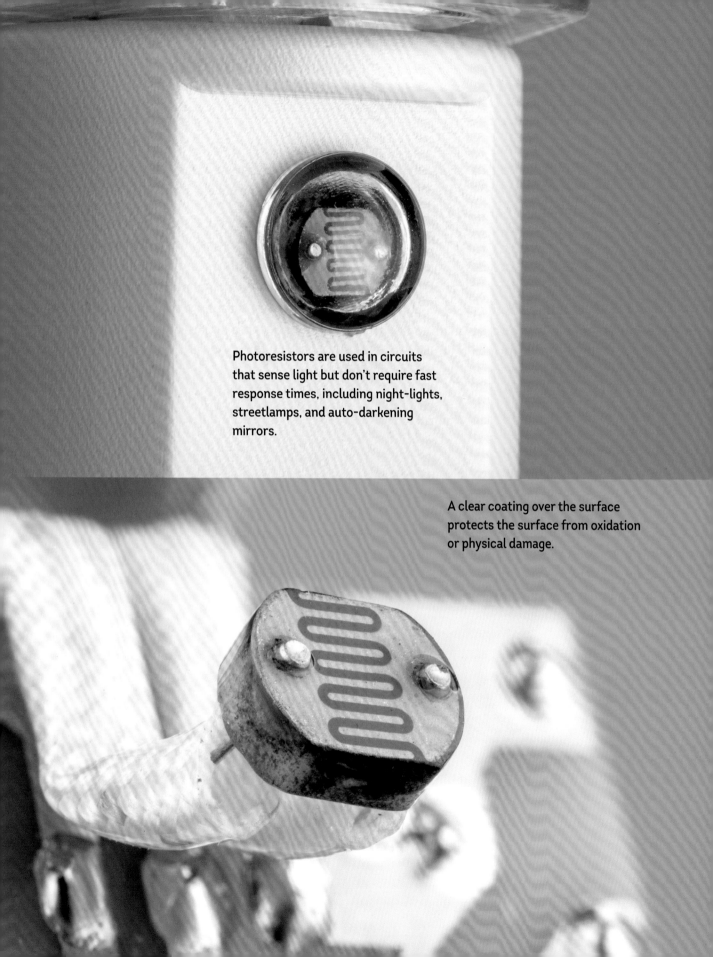

Photoresistors are used in circuits that sense light but don't require fast response times, including night-lights, streetlamps, and auto-darkening mirrors.

A clear coating over the surface protects the surface from oxidation or physical damage.

Point Contact Diode

A **POINT CONTACT DIODE**, sometimes called a **CAT WHISKER DIODE** or **CRYSTAL DIODE**, can be formed when a thin metal wire (the "whisker") contacts a chunk of semiconductor crystal.

This particular point contact diode has a steel wire that contacts a piece of galena, a naturally occurring lead sulfide mineral that happens to be a semiconductor. The wire position is adjustable to contact different points on the crystal, allowing the user to hunt for the best-performing areas on the surface.

A crude AM radio receiver can be built with nothing more than a point contact diode like this, a length of wire to form an antenna, and an earphone. No batteries are required; a **CRYSTAL RADIO** is directly powered by the radio waves that it receives.

Germanium Diode

The classic GERMANIUM DIODE can be found in crystal radio sets even today, but they've been largely superseded by more modern diode designs. This is a point contact diode and the semiconductor is a piece of germanium, rather than galena or the silicon found in other types of diodes.

The outward construction is very similar to the other glass-encapsulated diodes that we looked at on page 66. Inside, a tiny gray square of germanium is soldered to the copper cathode.

The "cat whisker" in this diode is a slender gold wire, formed into a spring shape and given an extremely fine point. The wire is welded to the anode lead and inserted into the glass envelope until it makes contact with the germanium. A current is then passed through the cat whisker and the germanium to fuse them together.

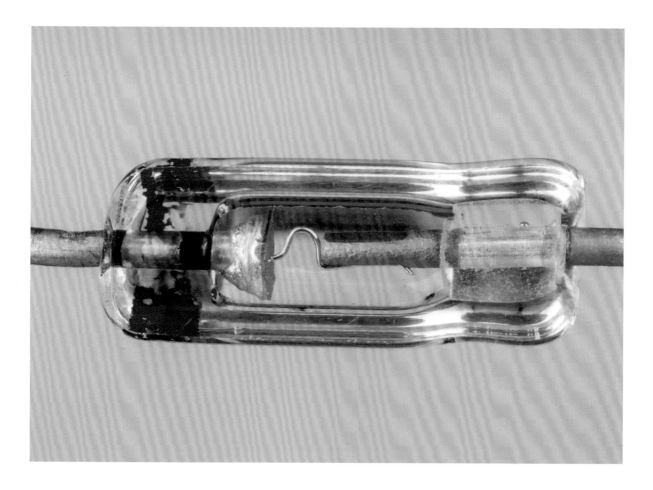

µA702 Integrated Circuit

This chip, marked µA702, was the first analog integrated circuit chip to reach the market. It was designed by legendary IC designer Bob Widlar of Fairchild Semiconductor and released in 1964. There are a grand total of nine transistors on the silicon die.

The µA702 is an **OPERATIONAL AMPLIFIER**, a device that subtracts and amplifies analog signals. Operational amplifiers are basic building blocks for analog circuits in much the same way that logic gates are for digital circuits.

Inside the type TO-99 metal can package, we can see evidence that this particular device was assembled or reworked by hand; it looks like the chip was placed off-center and then slid into place, leaving a streak of epoxy behind. The dark scratches on the right are probably marks from a tweezer, perhaps the one used to move the chip. Later on, chips were assembled with automated equipment and bear no evidence of human touch.

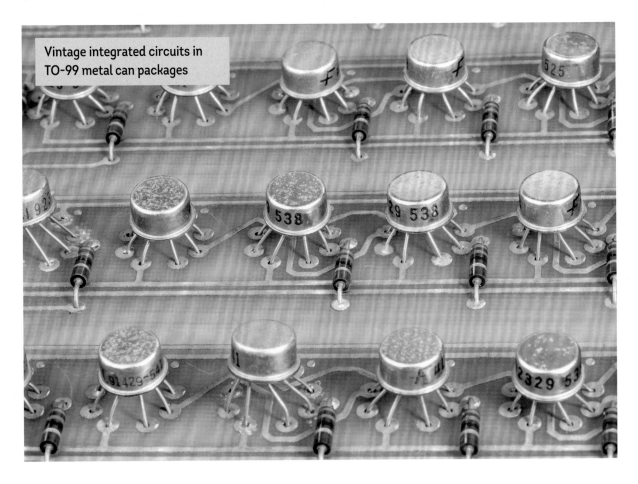

Vintage integrated circuits in TO-99 metal can packages

The chip is connected through eight
bond wires. Seven go to pins insulated
with glass seals. The eighth pin is
electrically connected to the can.

Windowed EPROM

READ-ONLY MEMORY, or **ROM**, is permanent computer memory that can only be read, not written. In practice, many ROM devices need to be programmed at some initial point. It's also convenient if devices can be erased and rewritten rather than being thrown away after one use.

This **ERASABLE PROGRAMMABLE READ-ONLY MEMORY (EPROM)** chip has a quartz-glass window that allows the silicon die to be exposed to ultraviolet light, which wipes the memory clean, resetting every bit to a digital 1. Later, the chip can be reprogrammed, setting individual bits to 0, to store data.

EPROM chips were once frequently used as BIOS chips on computer motherboards, usually with an opaque sticker covering the window. They have been superseded by **EEPROM**, *electrically* erasable programmable read-only memory, and its descendant, flash memory.

The ceramic DIP of an EPROM has
a quartz-glass window that allows
ultraviolet light to erase the memory
on the silicon die.

Core Memory

Before cheap memory chips, CORE MEMORY was one of the few reliable technologies used for a computer's main memory. Unlike a modern memory chip, which contains billions of bits, core memories contain so few bits that you can see every single one. Each donut-shaped core is made of a ferrite ceramic that can be magnetized in one of two directions, representing a binary 1 or 0.

The grid of red-colored horizontal and vertical wires passing through the cores is used to address an individual core for writing or reading. The diagonal copper-colored wire through every core is a SENSE LINE for reading back the information stored in the selected core. The black wire is an INHIBIT LINE that can selectively block information from being written to a set of cores by counteracting the magnetic field of the red wires.

The core memory plane from a Casio AL-1000 calculator, circa 1967, with 448 bits, or 56 bytes, of memory

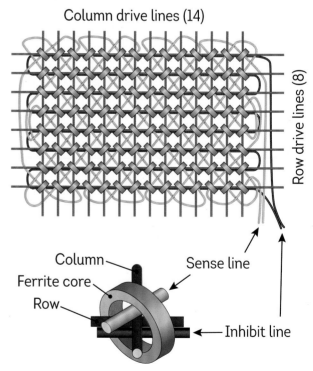

Column drive lines (14)

Row drive lines (8)

Column

Sense line

Ferrite core

Row

Inhibit line

IBM SLT Module

In the 1960s, IBM developed a type of hybrid circuit module that they called **SLT**, for Solid Logic Technology. These compact, rugged modules replaced an entire circuit card's worth of components, fitting resistors, diodes, and transistors into a package smaller than a sugar cube.

SLT modules contain multiple tiny chips, each with either a single transistor or an array of two diodes. Instead of using bond wires, solder bumps on the dies connect them to conductive silver circuits patterned on the module's ceramic substrate. Far ahead of its time, this solder-bump approach was the direct precursor of the "flip chip" mounting we saw in the smartphone SoC on page 87.

It almost seems anachronistic to see vintage components like carbon composition resistors on a circuit board next to these high-density SLT modules.

Individual SLT transistor dies. Notice the solder bumps on each die.

A bare ceramic SLT module, where the dies have not yet been bonded in place

SLT modules were packaged with an aluminum cover over the ceramic circuit board.

Analog Panel Meter

In the years before cheap LCD panels and LED displays, analog meters were used to indicate voltages and currents in a wide range of applications.

The type of analog meter shown here features a fixed permanent magnet that pushes against and turns an electromagnet in order to rotate the attached needle.

The rotating electromagnet is suspended by and pivots about two taut metal ribbons, one above and one below the coil. The ribbons conduct electricity from the meter terminals to the rotating coil. They also act as a weak torsion spring, returning the needle toward zero when the amount of current decreases.

The needle's angle of rotation is precisely determined by the current flowing in the electromagnet's copper coil, balanced by the weak spring. This mechanism is called the **D'ARSONVAL MOVEMENT**.

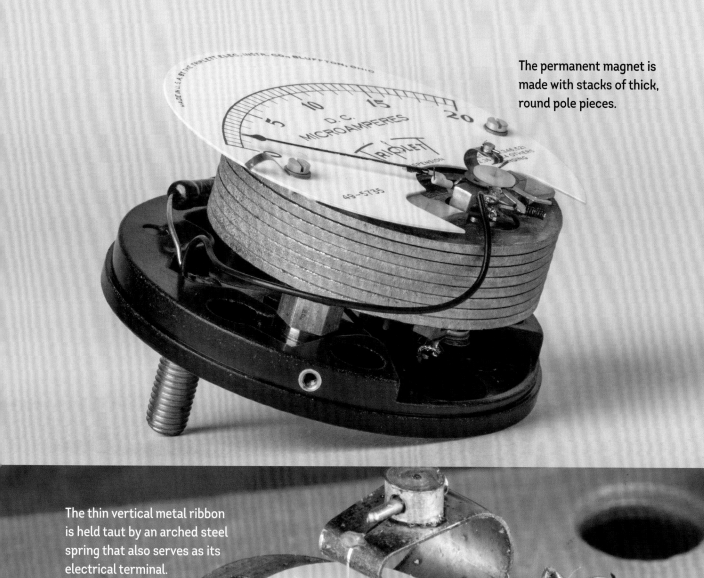

The permanent magnet is made with stacks of thick, round pole pieces.

The thin vertical metal ribbon is held taut by an arched steel spring that also serves as its electrical terminal.

Magnetic Tape Head

TAPE HEADS read or write information, like analog music or digital data, to or from magnetic tape. They may look simple from the outside, but their smooth exterior hides a complex assembly within.

A tape head's heart is a coil of copper wire. The coil, along with a shaped iron "C core" pole piece, acts as an electromagnet, concentrating a magnetic field at a tiny gap right at the point where the head presses up against the magnetic tape. The magnet and pole piece work much like a common horseshoe magnet: putting the two poles close together creates a strong magnetic field across the gap.

The gap itself is formed by a thin piece of copper or gold foil sandwiched between the two ends of the iron core. The foil ensures that the gap is of a precisely controlled and consistent width, giving better overall fidelity and performance.

Two tape heads are visible in a Panasonic-brand cassette tape recorder. The metal one in the center is for playback and recording. The white plastic head beside it is for erasing a preexisting recording.

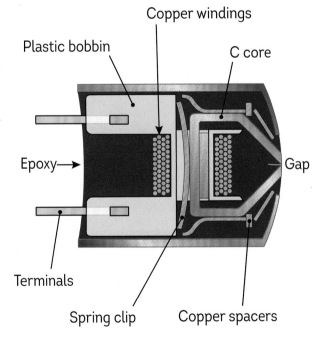

Copper elements are used to separate and position the iron pole piece within the head, isolating the magnetic circuit components.

The thin foil is pinched between the two ends of the ferromagnetic core, visible as a light, wispy line against the black epoxy.

Thin-Film Hard Drive Head

A computer hard drive contains a miniature version of a magnetic tape head. More precisely, it contains many of them.

The heads sit on ceramic **SLIDERS** that are positioned in parallel by a rotary voice coil motor. Every disk-shaped platter of the drive is paired with two sliders: one for each surface. When the drive spins, each slider glides on an ultra-thin cushion of air between it and the surface.

Each individual slider is quite small, only about 3.3 mm wide. On its front face are metal terminal pads and two tiny, ruby-colored heads. The heads are manufactured using thin-film techniques to deposit coils that are electrically similar to tape heads.

Only one head per slider is actually used. They're given two heads so they can be manufactured identically, even though the slider underneath each platter is flipped over. For these upside-down sliders, the opposite head is used, lining up the heads above and below the platter so that each data track has the same diameter.

This 2 GB Micropolis-brand hard drive from 1992 has eight platters for storing data.

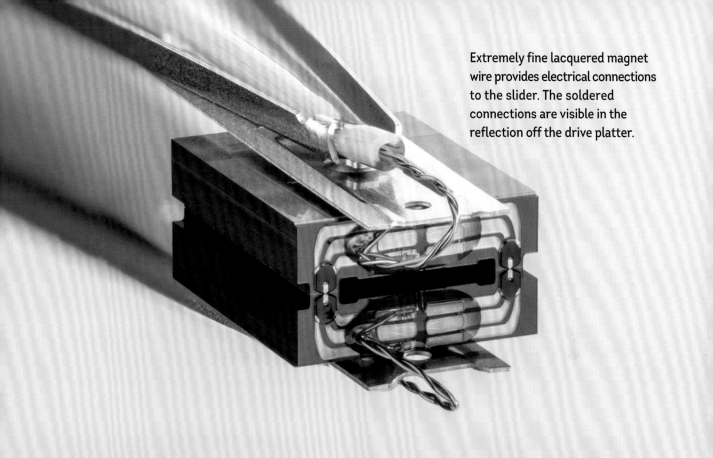

Extremely fine lacquered magnet wire provides electrical connections to the slider. The soldered connections are visible in the reflection off the drive platter.

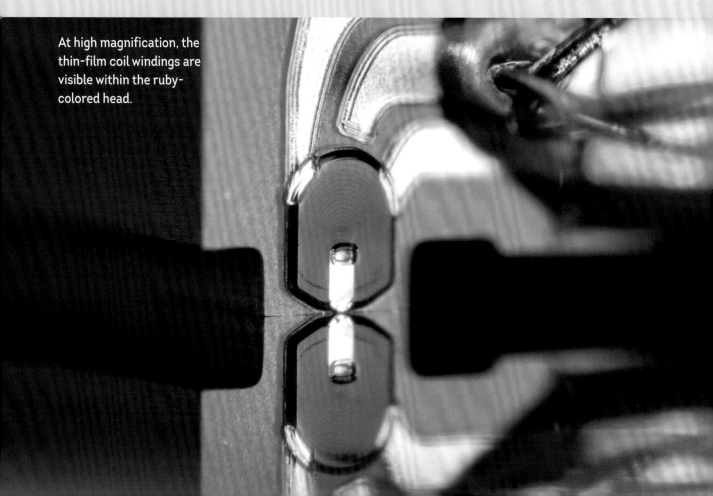

At high magnification, the thin-film coil windings are visible within the ruby-colored head.

GMR Hard Drive Head

Compare the thin-film hard drive heads to this hard drive from 2001. By this time, heads were using **GIANT MAGNETO-RESISTANCE (GMR)** technology, and the sliders had shrunk to only about 1 mm wide.

Magnetoresistance is a phenomenon where the resistance of a material can change in the presence of an external magnetic field. Sensors based on this effect can detect very small magnetic domains, allowing drive memory storage density to increase.

As a *sensor* technology, a GMR-based head doesn't on its own write data to the drive. Rather, a GMR "read head" is layered with a thin-film "write head" like the one that we saw previously.

The hard drive industry continues to evolve. Current-generation drive heads use entirely different technologies to achieve even greater storage density.

This 100 GB Western Digital hard drive has only three platters. A picture of the full drive can be found with the rotary voice coil on page 132.

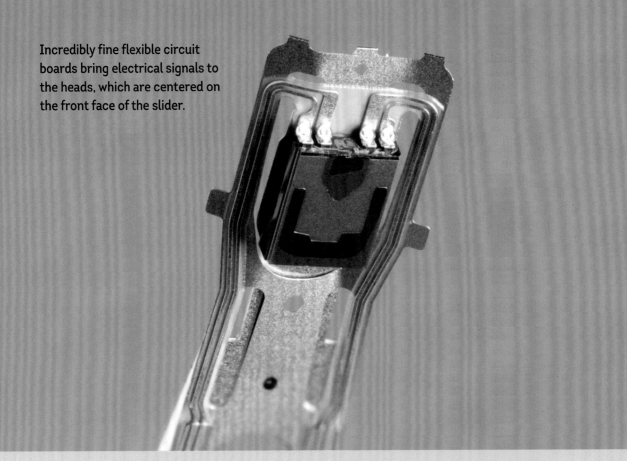

Incredibly fine flexible circuit boards bring electrical signals to the heads, which are centered on the front face of the slider.

The red thin-film write head coil is at the center bottom of the slider face. The two copper rectangles behind it are part of the GMR sensor read head.

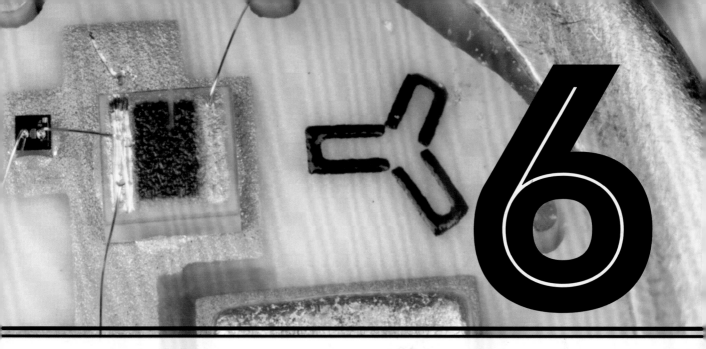

6

Composite Devices

Take apart some electronic components, and you'll find other, smaller components inside them. And inside those components? Yes, sometimes, you'll find even smaller components inside those as well. We now turn our attention to some of these composite devices. We'll see all manner of circuit boards, displays containing multiple LED dies, composite packages that combine multiple components, and exotic hybrid modules with ceramic circuit boards and individual semiconductor dies linked by delicate bond wires.

LED Filament Light Bulb

Modern LED light bulbs are available in a wide variety of shapes and styles. This one is designed to look like an old-school incandescent light bulb. But how can LEDs be made so long and thin?

Each "filament" is actually a ceramic strip—essentially a circuit board—with dozens of tiny blue LED dies studded along its length. Each ceramic strip is coated on its front and back sides with a yellow silicone rubber filled with a phosphor.

As with other "white" LEDs, the phosphor absorbs some of the blue light and emits a broad spectrum of light that stretches into the green and red. The overall light that we perceive is a warm white glow, like that from an incandescent bulb.

LED filament bulbs have much of the aesthetic appeal of incandescents, while converting electricity to light more efficiently.

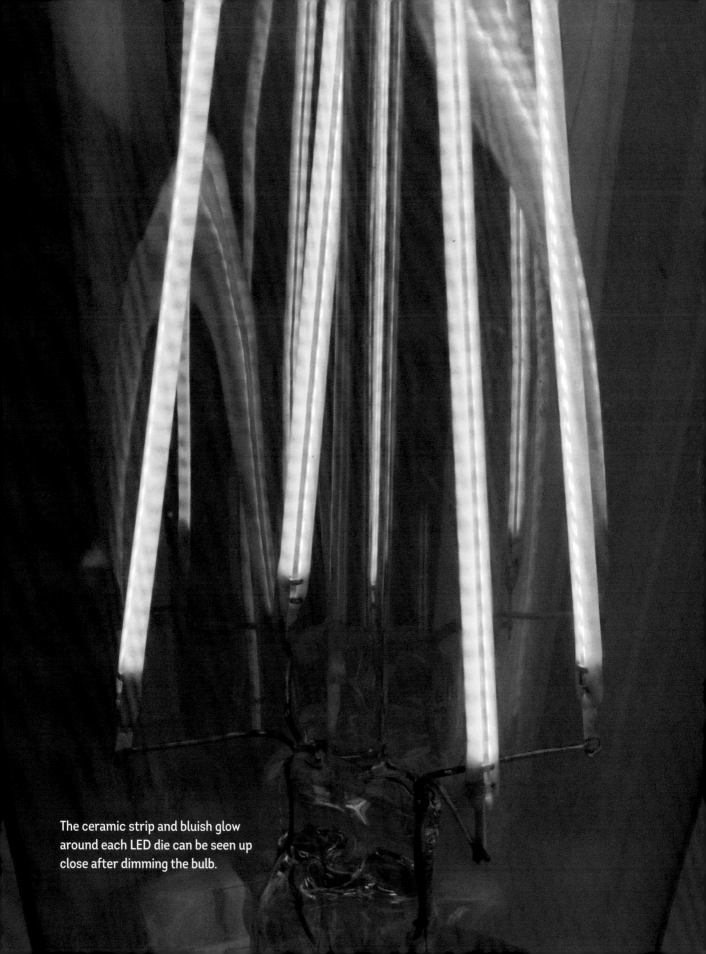

The ceramic strip and bluish glow around each LED die can be seen up close after dimming the bulb.

Single-Side Printed Circuit Board

PRINTED CIRCUIT BOARDS, or PCBs, are everywhere in electronic devices. Despite the name, the copper circuitry isn't printed on the board. Rather, the board—usually a fiberglass composite—is bonded to a sheet of copper and selectively etched, leaving only the desired wiring patterns on the board, called TRACES. Holes for component leads are drilled through the board as well.

The circuit board shown here, part of a power supply, is called a *single-sided* board because it has copper traces on one side only. The side without copper has a number of through-hole components, including a DIP chip, a transistor, and film capacitors. The reverse side, where we can see dark, snaking lines of copper, has surface-mount components, including chip resistors and chip capacitors.

The green coating on the copper side is a thin insulating layer called a SOLDER MASK. Solder won't stick to areas covered by this coating.

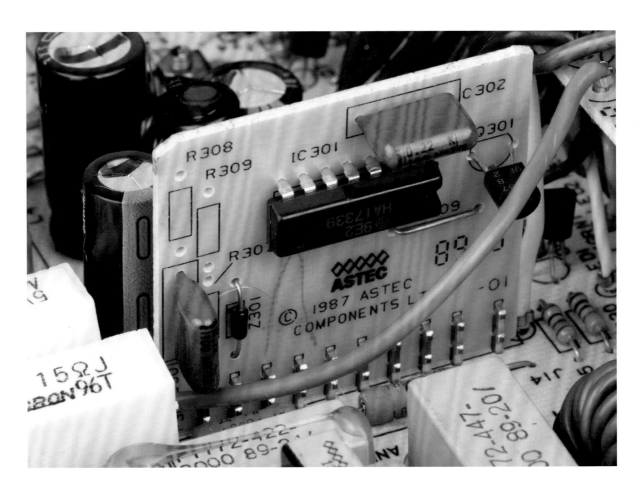

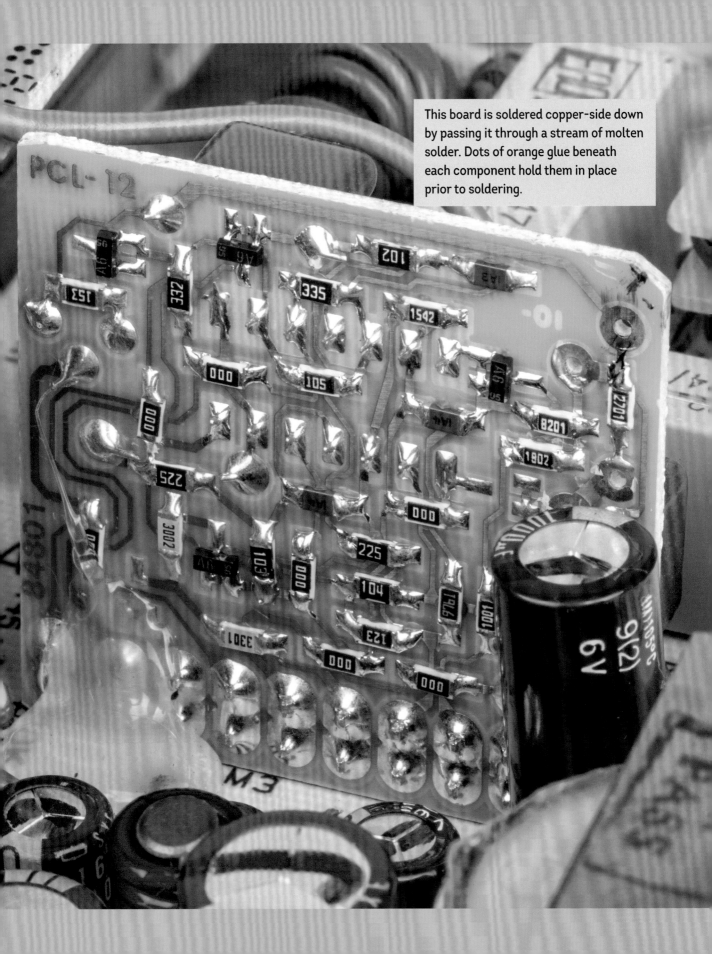

This board is soldered copper-side down by passing it through a stream of molten solder. Dots of orange glue beneath each component hold them in place prior to soldering.

Two-Layer Printed Circuit Board

While single-sided PCBs are cheap to manufacture, they're notoriously difficult to design because wiring traces can't cross over each other. By bonding a sheet of copper to each side of a fiberglass substrate and etching circuits into both sides, traces on one side can then pass over traces on the other side. It's much easier to plan the routes for wiring on a two-layer circuit board.

Copper traces on the two layers can be connected with **PLATED THROUGH HOLES**. After a hole is drilled, a chemical process electroplates additional copper inside the hole, connecting the two sides. Plated through holes added to connect traces between layers are called **VIAS**. Other plated through holes are used as locations for soldering components.

This circuit board has a purple solder mask and a thin gold plating on the exposed copper surfaces.

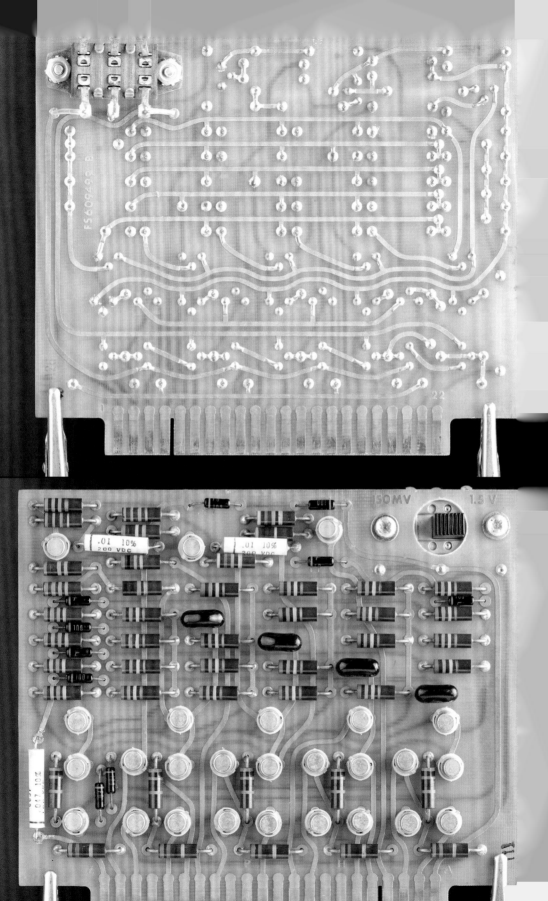

Multilayer Circuit Boards

Densely populated circuit boards often need more than just two layers for routing their wiring.

A manufacturing plant can build PCBs with more than two layers by creating more complex sandwiches of copper and fiberglass. A four-layer board is typically fabricated by etching two thin two-layer boards, and then laminating them to either side of a common fiberglass core. The laminating process permanently bonds the layers together in a high-pressure, high-temperature press. The same process, but with different layer choices, can produce circuit boards with dozens of layers.

Special features like blind vias and buried vias—plated vias that don't go through the entire circuit board—can be fabricated by drilling and plating holes in the various layers before they're laminated together.

A four-layer circuit board

A six-layer circuit board

This ten-layer smartphone PCB has blind and buried vias making connections between the different layers, visible as vertical copper pillars connecting the layers.

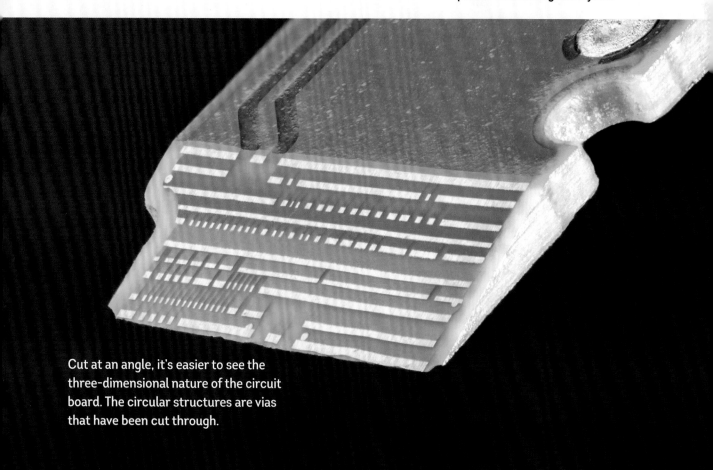

Cut at an angle, it's easier to see the three-dimensional nature of the circuit board. The circular structures are vias that have been cut through.

Flex and Rigid-Flex Circuit Boards

FLEX PCBs are circuit boards that are etched on a bendable substrate of polyimide film rather than fiberglass. Polyimide is a remarkably tough and flexible plastic that can withstand the high heat of soldering. It has a distinctive rich brown color and is frequently referred to by the brand names Pyralux or Kapton.

To create structure, flex PCBs are frequently manufactured with **STIFFENERS**, flat layers of fiberglass or other materials that are cut to shape and bonded to the polyimide layers. These help the board hold its shape in certain places—for example, where components are soldered on.

A **RIGID-FLEX PCB** is a true multilayer PCB where at least one of the layers in the stack is etched on a flexible polyimide substrate. These remarkable boards can behave like full-featured circuit boards with built-in hinges and cabling.

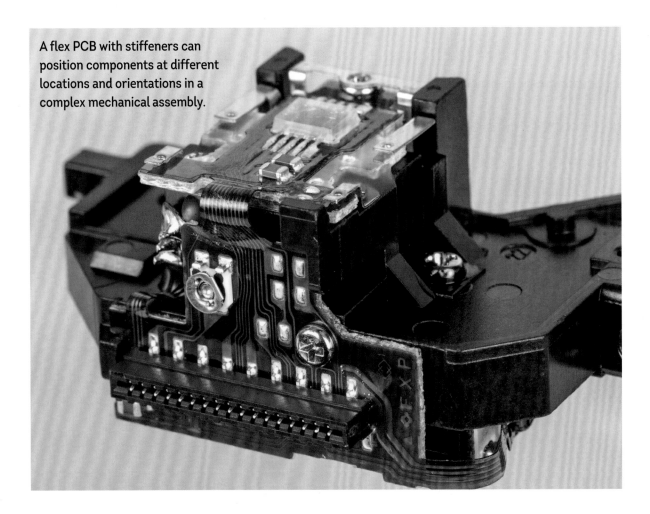

A flex PCB with stiffeners can position components at different locations and orientations in a complex mechanical assembly.

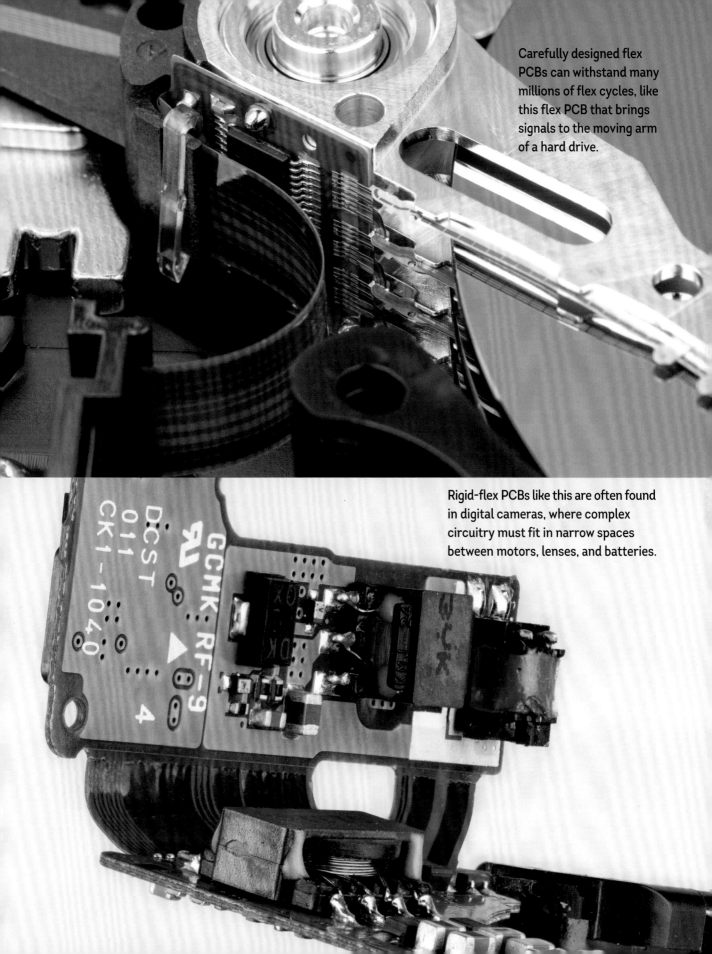

Carefully designed flex PCBs can withstand many millions of flex cycles, like this flex PCB that brings signals to the moving arm of a hard drive.

Rigid-flex PCBs like this are often found in digital cameras, where complex circuitry must fit in narrow spaces between motors, lenses, and batteries.

Elastomeric Connector

ELASTOMERIC CONNECTORS are a very different type of flexible circuit. They're commonly found in the liquid crystal display (LCD) modules of inexpensive devices like wristwatches and calculators.

A typical LCD module features a piece of glass patterned with transparent conductive electrodes, and a circuit board beneath to drive the display. The elastomeric connector is a squishy rubber strip that sits between the electrodes and corresponding contact pads on the circuit board, providing a reliable and gently compliant connection between the two.

The connector is composed of alternating parallel layers of insulating white silicone rubber and conductive carbon-filled black silicone rubber. There are also thin insulating white silicone liners on the top and bottom.

This electronics module from a digital wristwatch has an LCD connected to the circuit board by two elastomeric connector strips.

Two rows of gold-plated contact pads at the top and bottom of the circuit board correspond to the electrodes on the LCD module.

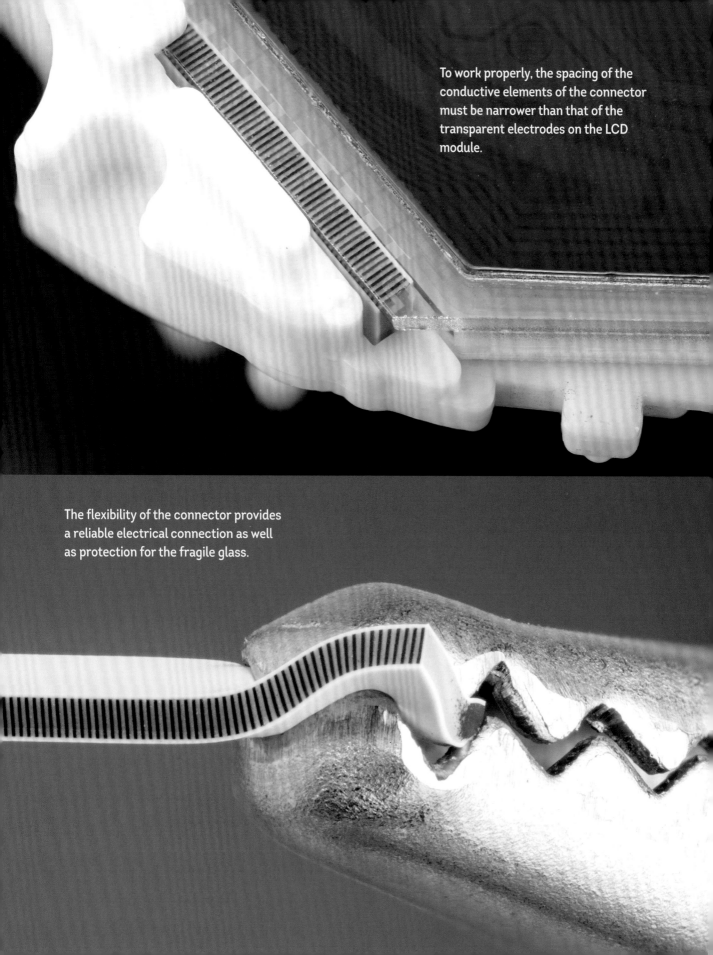

To work properly, the spacing of the conductive elements of the connector must be narrower than that of the transparent electrodes on the LCD module.

The flexibility of the connector provides a reliable electrical connection as well as protection for the fragile glass.

MicroSD Card

A microSD (Secure Digital) memory card contains a very thin circuit board with a memory chip. Since the whole device is encapsulated in black epoxy resin, it isn't immediately obvious that there's a circuit board involved—at least until the card is cut in half.

The memory chip looks like a silver gray strip. Surprisingly, it fills most of the inside of the card, even underneath the gold-plated connector fingers. Making efficient use of space allows manufacturers to maximize the amount of storage. This is the polar opposite of simple transistors like the 2N2222 (see page 70), where the package is far larger than the active silicon. It's also a natural result of 40 years of evolutionary change in electronics manufacturing and packaging.

MicroSD cards are about the size of a dime, yet as of this writing they're available with capacities up to 1 TB.

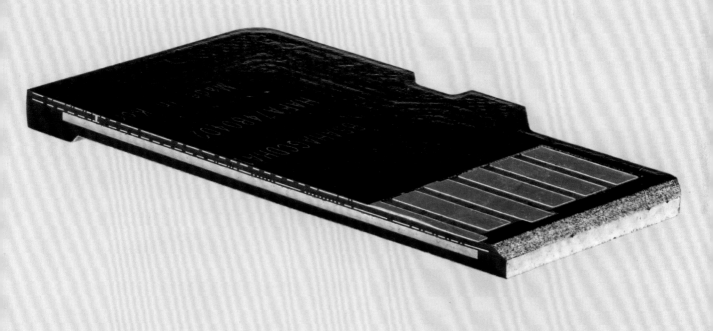

The entire back surface of the microSD card is a two-layer circuit board. The large contact fingers are simply exposed copper areas from the upper layer of the PCB, with a gold-plated finish.

Glob-Top Packaging

Ultra-low-cost electronics, like dollar store calculators or unbranded multimeters, need to save every cent possible in the manufacturing process. Products like these often don't use chips in a traditional epoxy package with metal pins that need to be soldered in place. To save money, the IC die itself is glued to the circuit board and connected to the circuitry using very fine bond wires. A glob of epoxy covers the fragile wires, and the chip is ready for use.

This technique is called GLOB-TOP PACKAGING, and we've already seen an example of it in the wristwatch on page 236.

Glob-top packaging is a form of CHIP-ON-BOARD, or COB, packaging, where dies are attached directly to the circuit board. COB packaging is commonly used in LED lighting; the "filament" modules in the LED filament light bulb (page 226) are built this way.

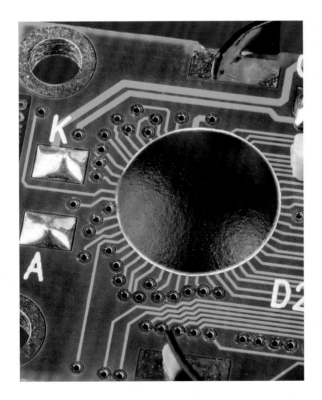

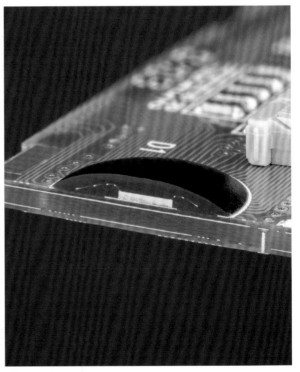

Several aluminum bond wires can be seen here, where the chip is connected to the circuit board.

EMV Credit Card Chip

Modern credit cards come embedded with a secure memory EMV CHIP instead of just a simple magnetic stripe. The name stands for Europay, Mastercard, and Visa, the companies that created the standard.

Dissolving away the rest of the credit card in a solvent, we can see that the visible contact pads fill one side of a paper-thin circuit board embedded in the card.

The rectangular circuit board is made out of fiberglass and epoxy. The memory chip is glued to the back side, connected to the contact pads with hair-thin gold bond wires, and protected with a glob of clear epoxy. It's another example of glob-top packaging.

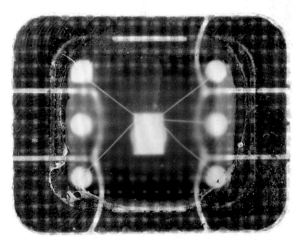

Different credit cards and systems use different contact pad patterns. These two examples have slightly different contacts.

NFC Key Card

Modern hotel doors use **NEAR-FIELD COMMUNICATION (NFC)** technology to identify key cards that have permission to unlock the door. Molded inside each card is a coil of wire connected to a tiny smart card integrated circuit.

When the card is tapped against a door lock, electronics in the lock interrogate the card using a modulated magnetic field generated by a coil inside the lock. This same magnetic field also provides electrical power to the smart card chip. Another way to look at it is that bringing the card up to the lock creates an ad-hoc transformer, with one set of windings in the lock and the other in the card.

The chip is on a paper-thin circuit board, much like that in the credit card, complete with the black glob-top over the IC. Interestingly, the circuit board is intentionally bent with a slight curve.

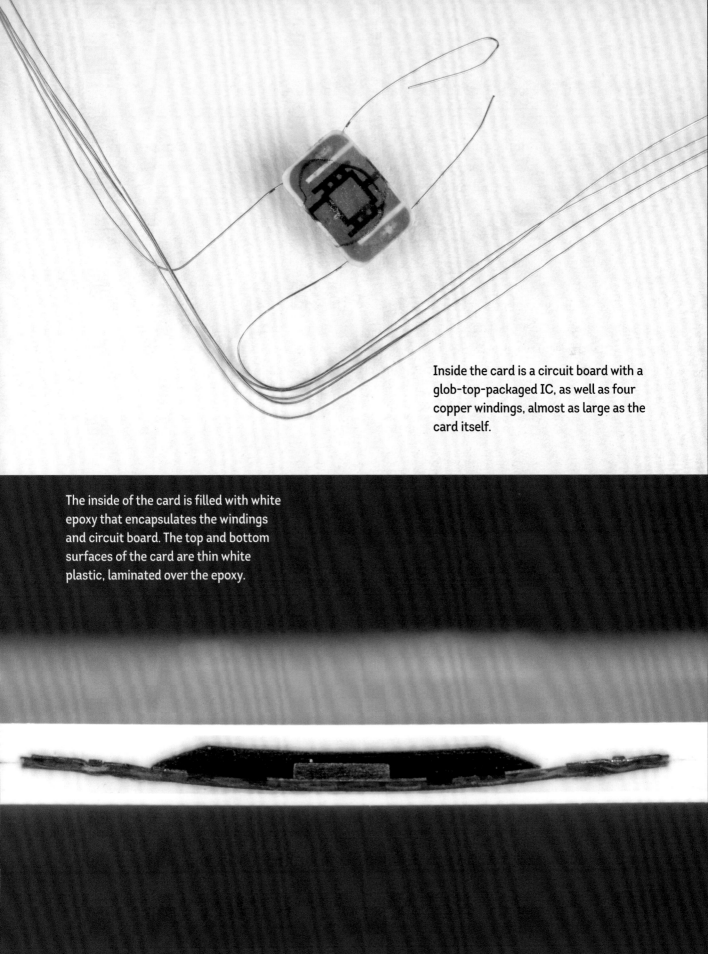

Inside the card is a circuit board with a glob-top-packaged IC, as well as four copper windings, almost as large as the card itself.

The inside of the card is filled with white epoxy that encapsulates the windings and circuit board. The top and bottom surfaces of the card are thin white plastic, laminated over the epoxy.

Smartphone Logic Board

We often take our smartphones for granted, but they're technological marvels teeming with a remarkable array of circuitry and sensors.

Inside this smartphone is a printed circuit board with a blue solder mask, the same one that we saw cross-sectioned on page 233. Engineers crafted it to fit the mechanical peculiarities of this particular phone, cutting out a large region for the lithium polymer battery pack and

smaller regions for the camera module and various mounting features.

Minuscule components encrust the surface of the board, including connectors, crystals, the LED camera flash, and various tiny sensor modules, such as a microphone and accelerometer. Most of the integrated circuits lurk beneath thin metal covers, which clip into place to provide shielding for the sensitive electronics within.

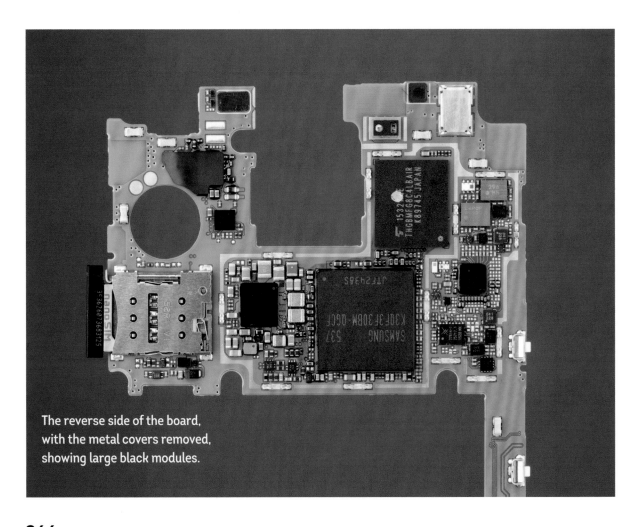

The reverse side of the board, with the metal covers removed, showing large black modules.

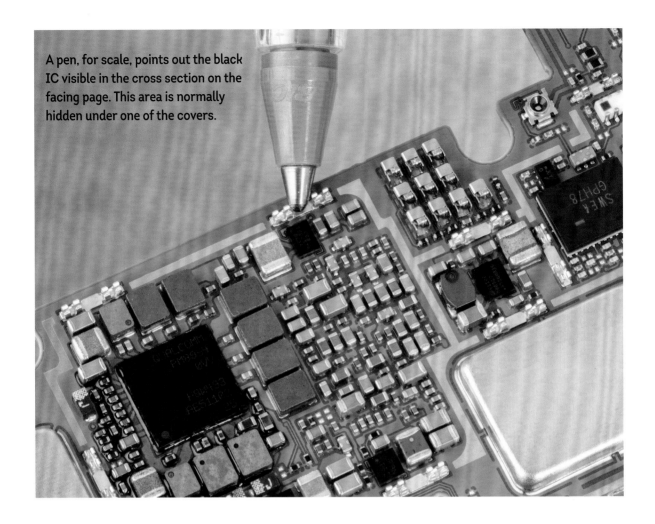

A pen, for scale, points out the black IC visible in the cross section on the facing page. This area is normally hidden under one of the covers.

Inside the Logic Board

Circuitry can be a profoundly three-dimensional art form. A slice through the center of the smartphone logic board reveals a thick sandwich of materials, all part of different components.

The top layer is the blue main printed circuit board, containing 10 layers of copper wiring. Beneath it, connected to the board through a ball grid array, is a complex SYSTEM IN PACKAGE (SiP), a component containing multiple ICs. The

SiP starts with a PCB made of six layers of copper circuitry. A large, thin chip is soldered to this six-layer board using microscopic solder bumps.

Lower down on the SiP is another PCB, made of three layers of copper circuitry. Mounted to that board and connected with fine gold bond wires are at least two additional chips. The entire SiP assembly is encapsulated in black epoxy.

The brown blocks atop the main circuit board are multilayer ceramic capacitors (MLCCs).

Ethernet Transformer

Networking cable connections need to be electrically isolated for safety reasons. Computers and other devices that have Ethernet ports use a group of toroidal transformers (page 50) to achieve that isolation.

Cutting into this **ETHERNET TRANSFORMER** reveals eight tiny toroidal transformers at various angles, two for each of the four twisted wire pairs in an Ethernet cable (page 162). One transformer in a pair provides electrical isolation, while the other is configured as a choke to filter out noise common to both wires.

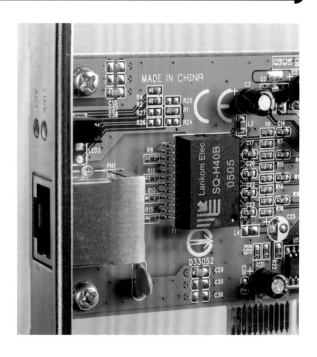

DC-DC Converter

A voltage regulator like the LM309K (page 74) is reliable but power hungry. It lowers and controls a voltage by converting electrical energy to heat.

A **DC-DC CONVERTER** also changes one voltage to another, but does so with much greater efficiency. It uses a digital circuit to manage the current through an inductor, exploiting its flywheel-like behavior.

This miniature DC-DC converter module is meant to replace a similarly shaped linear voltage regulator in existing equipment, improving its efficiency.

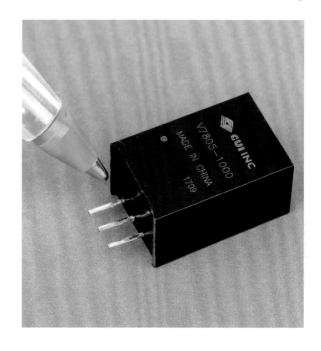

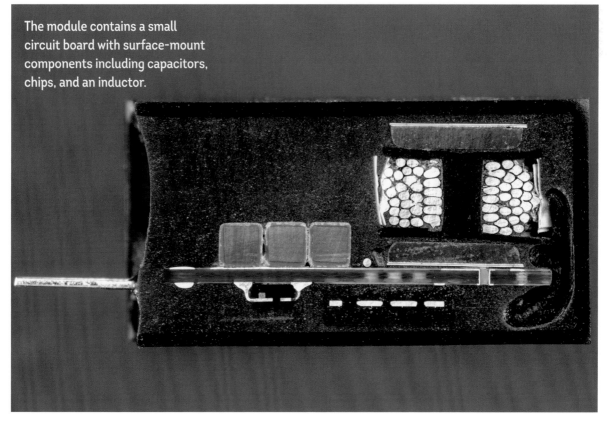

The module contains a small circuit board with surface-mount components including capacitors, chips, and an inductor.

Seven-Segment LED Display

One surprising thing about a SEVEN-SEGMENT LED DISPLAY is how tiny the actual LEDs are in comparison to the overall device. At the base of each D-shaped colored plastic lens sits a small LED chip. These are about the same size as the LED dies in the red through-hole LED (page 88), but the packaging here is much larger.

The LED dies are mounted chip-on-board style to a small circuit board. Fine bond wires make connections from the dies to the copper traces on the board, which bring the drive signals in from the large metal pins. The board itself is single-sided, made out of a black fiberglass substrate to minimize reflections.

The circuit board assembly is placed into the white outer frame, made from injection-molded plastic with black paint on the front surface. Finally, the assembly is filled with red-tinted epoxy resin that hardens into the visible lenses of the device.

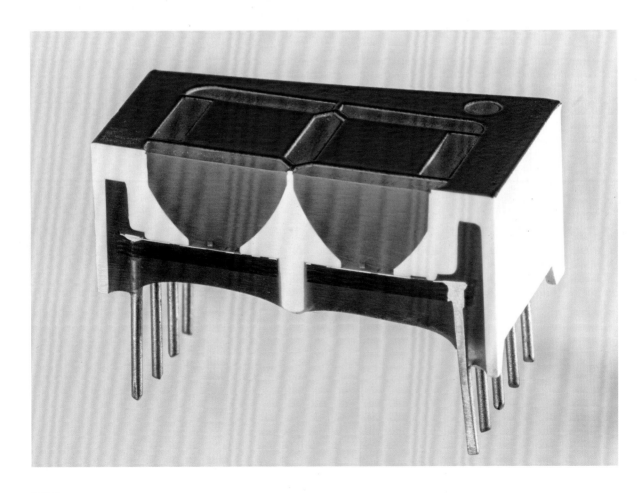

The slightly cloudy red epoxy that forms the lenses diffuses the light from the LEDs so that each lit segment appears evenly illuminated.

Thick-Film LED Numeric Display

This HDSP-0760 LED display is a high-end alternative to the common seven-segment LED display. Instead of concealing a single LED in each of seven lenses, the device features a pattern of 20 directly visible LED dies.

The LED display is a thick-film hybrid circuit with a ceramic base—there's no plastic involved at all, unlike the seven-segment LED display. Like the other thick-film devices that we've seen, the display's fabrication involves multiple steps of printing materials like gold traces and ceramic inks, with intermediate firing steps. A glass cover seals the completed assembly so that it can perform well in harsh environments that would melt or destroy other types of displays.

The chip inside this display converts the incoming binary number into the correct pattern of lit and unlit LEDs to form the matching character to display.

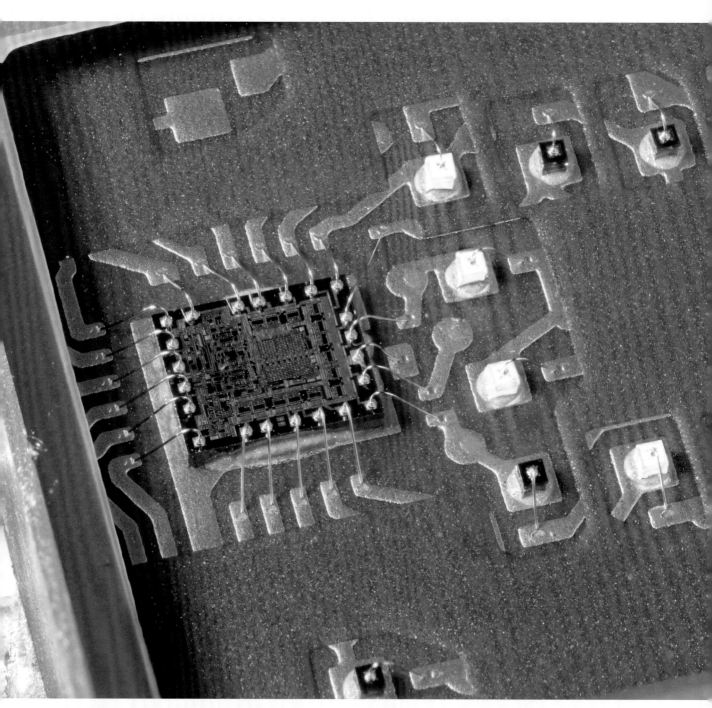

The glass cover of the display provides an exceptional view of lit and unlit LED dies, as well as the decoder chip and its bond wires.

5×7 LED Dot Matrix Display

Instead of constructing a small range of characters with a few dots or segments, this HCMS-2904 display uses 5×7 grids of individual LED dies to form any alphanumeric character.

The display is fabricated as a regular multilayer circuit board, with 140 LED dies mounted to the top, chip-on-board style. The dies and their bond wires are protected by a clear plastic cover over

the top of the device, a low-cost alternative to the thick-film LED display's glass-ceramic package.

These dot matrix modules are designed to stack seamlessly end to end and top to bottom to create larger displays. Each module contains a driver chip on the underside of its circuit board. The chip accepts a stream of pixel data and manages the LED display.

The LED dies are arranged in a matrix, with circuit board traces defining the columns in this view and daisy-chained bond wires defining the rows.

Vintage LED Bubble Display

Early electronic calculators like this HP-67 used seven-segment LED displays instead of the now ubiquitous LCD panel. Each digit is a single LED chip with seven segments and a decimal point patterned on it. Because each LED chip is small, magnifying lenses molded into the outer plastic case of the display enlarge the digits, making them easier to see.

In the close-up photo of the digits, you may be able to see the array of fine gold bond wires that connect the LED digits to the rest of the calculator's circuitry. To reduce the number of pins, a technique called MULTIPLEXING is used. Each group of five digits shares the same control pins for each segment—for example, all five digits' "top" segments are wired together. Because of the shared wiring, only one digit can be lit at a time. The circuitry cycles quickly through all the digits, making it appear like each digit is continuously lit.

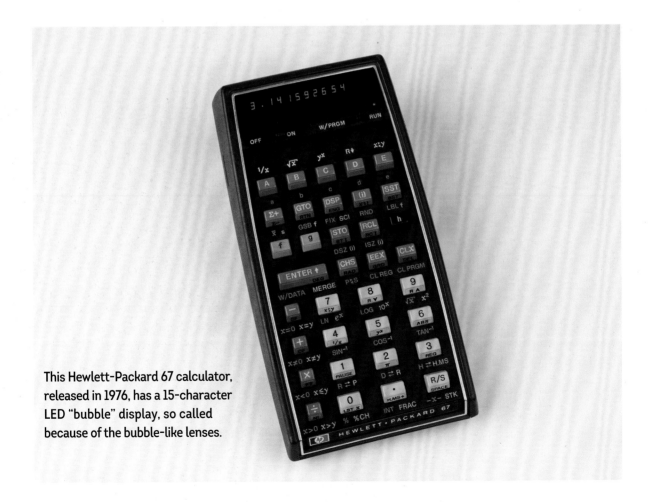

This Hewlett-Packard 67 calculator, released in 1976, has a 15-character LED "bubble" display, so called because of the bubble-like lenses.

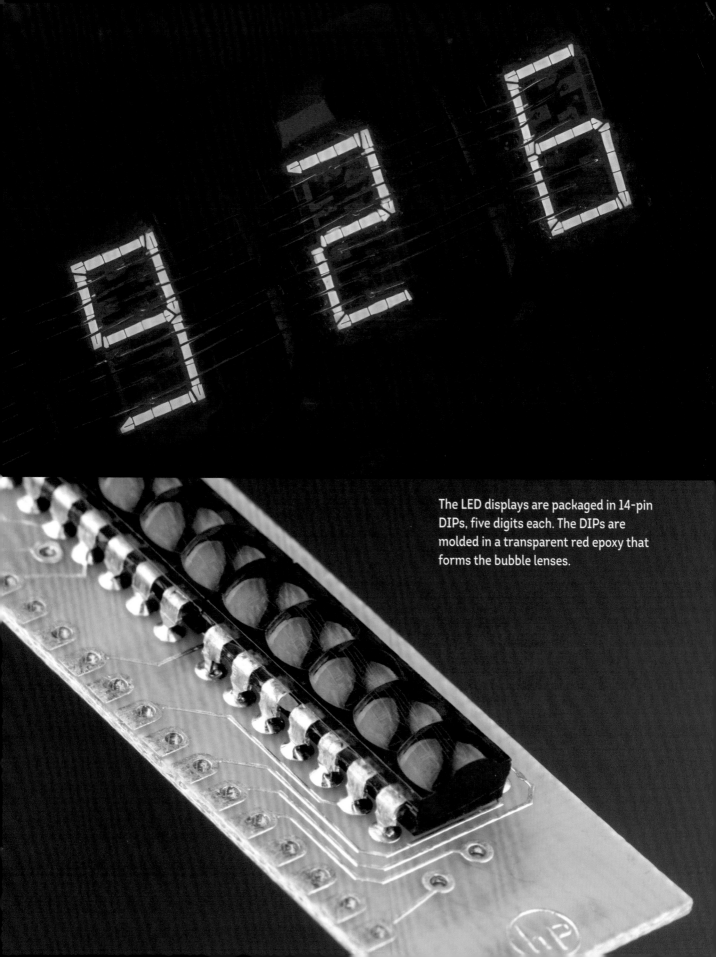

The LED displays are packaged in 14-pin DIPs, five digits each. The DIPs are molded in a transparent red epoxy that forms the bubble lenses.

Alphanumeric LED Display

While this display has similarities to the HP bubble display, its pedigree isn't from consumer electronics but rather from military and aerospace applications. It's a ruggedly built LED display, designed for cost-insensitive systems that require a tough, hermetically sealed display that just has to work.

The thick-film module is built with a glass cover above a ceramic substrate. On top of the ceramic are several layers of conductive silver traces. Insulating material is also patterned to allow traces to cross over each other without shorting out. The large 16-segment LED dies (17 with decimal) are wired to the traces with bond wires.

A chip on the opposite side of the ceramic decodes binary information representing letters and numbers into signals that drive the LED segments.

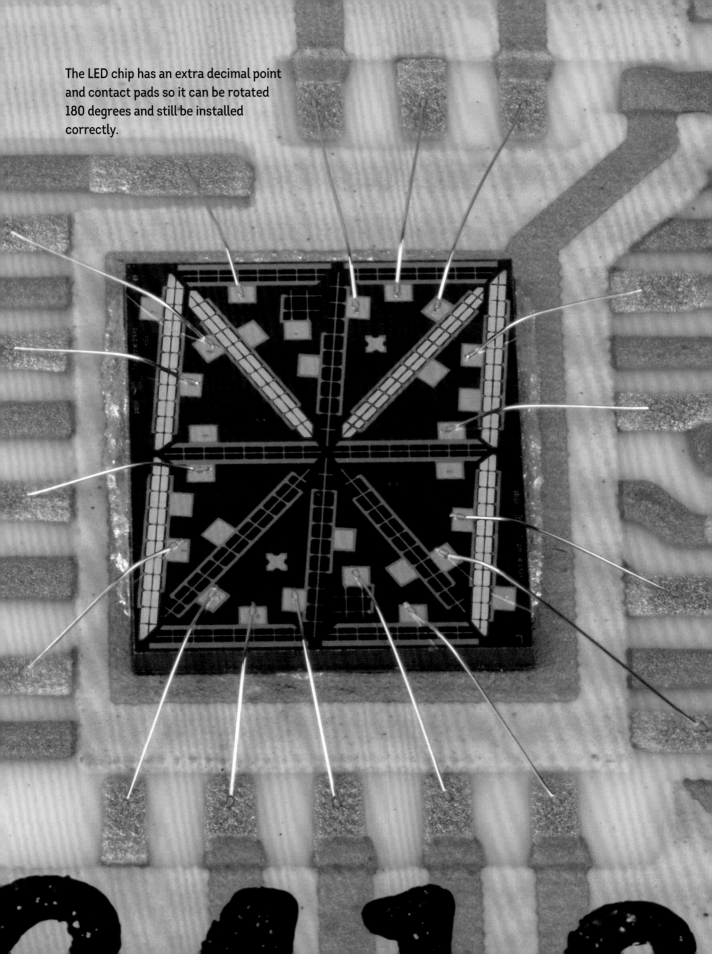

The LED chip has an extra decimal point and contact pads so it can be rotated 180 degrees and still be installed correctly.

Temperature-Compensated Clock

The DS3231 shown here is a hybrid circuit **REAL TIME CLOCK**, or **RTC**. A "clock" in electronics usually means an oscillating signal used to synchronize logic. By contrast, an RTC counts hours, minutes, and seconds as they elapse. It's a digital watch designed to be read by a computer.

From the outside, the DS3231 looks like a regular IC in a 16-pin SOIC package.

Inside, it has not only a chip but also a 32 kHz quartz crystal, molded right into the case. The crystal is nearly identical to that in the wristwatch on page 14.

The exact frequency of a quartz crystal changes with temperature, but a sensor on the chip compensates for these variations automatically. This arrangement is called a **TEMPERATURE-COMPENSATED CRYSTAL OSCILLATOR**, abbreviated **TCXO**.

The gray integrated circuit die can be seen on the left, atop the copper lead frame, leaving room for the 32 kHz quartz crystal on the right.

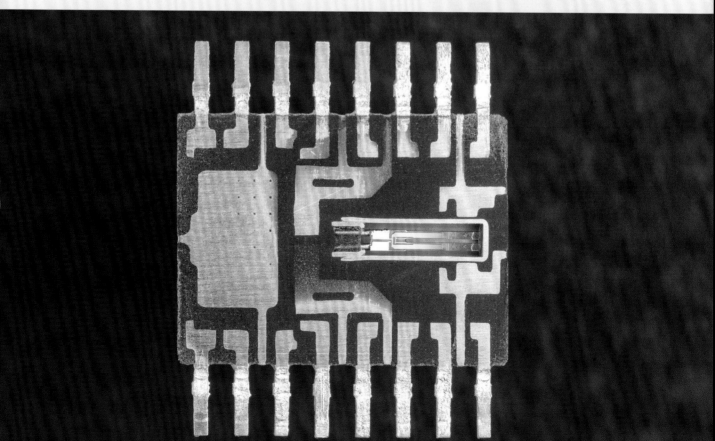

Crystal Oscillator

The "ticking heartbeat" of many digital devices comes from an oscillator module like this one. Inside, we find a paper-thin disc of sawn quartz crystal suspended on springs.

The disc is selectively plated with silver electrodes on its top and bottom sides. When voltage is applied to the electrodes, it stimulates motion in the quartz, turning it into a tiny, electrically driven clock pendulum. Unlike the tuning fork–shaped 32 kHz quartz crystal on page 15, the disc's oscillation frequency is much higher, perhaps 50 MHz. Discs are one of several geometries of quartz crystals used to achieve higher-frequency oscillations.

The module also contains a thick-film ceramic circuit board. On the board, a small surface-mount chip and capacitor compose the rest of the oscillator circuit, providing a clean square wave to the output pin.

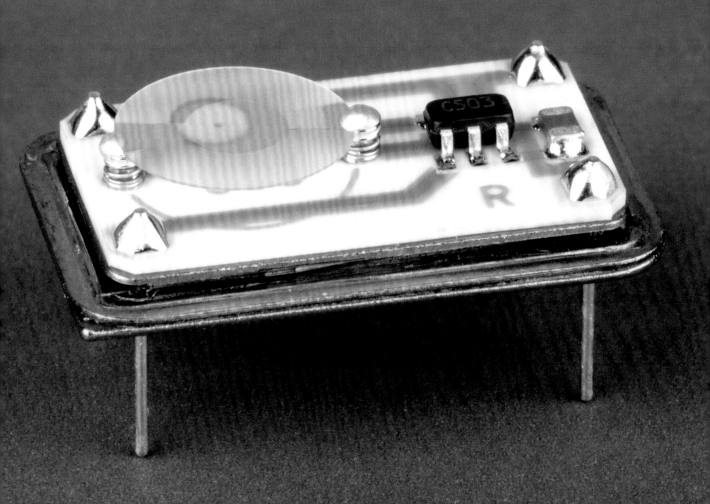

Tiny metal springs mechanically support and isolate the quartz. They also provide the electrical connection between the electrodes and the ceramic board below.

Avalanche Photodiode Module

This gold-plated AVALANCHE PHOTODIODE, or APD, MODULE has much higher performance than common photodiodes, and a corresponding price tag. It's a fast, high-sensitivity light detector with low noise characteristics, used in precision optical equipment for the communications industry as well as for scientific applications.

The module is packaged in a metal can with a glass top. The avalanche photodiode die is the gold square in the center of the hybrid module. Circuitry around the photodiode amplifies the tiny signals it generates.

If you look carefully, you can see surface-mount chip capacitors, diodes, transistors, and laser-trimmed thick-film resistors, all wired together with printed thick-film traces and nearly microscopic gold bond wires.

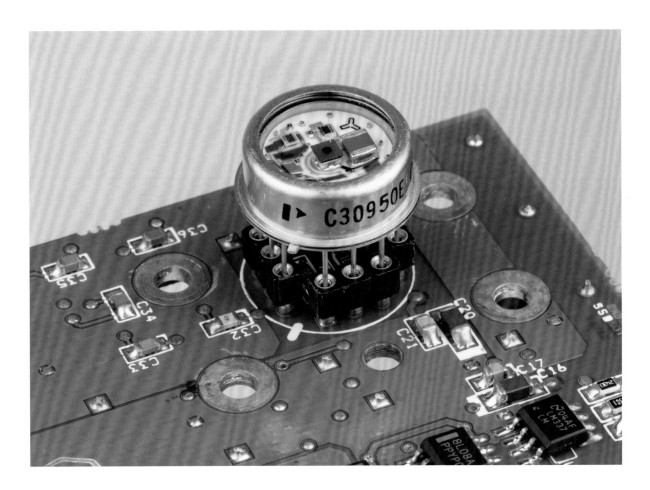

The triangular black symbol is the logo of EG&G, the original manufacturer of this device. It's now made by Excelitas.

3656HG Isolation Amplifier

Outwardly gray, this ISOLATION AMPLIFIER module turns out to be one of the most colorful and intricate of electronic components. It's a specialty part, designed for the medical and nuclear industries, which require some circuits to be electrically isolated for safety purposes. For example, a sensor circuit may need to operate at high voltage with respect to a computer that controls it. The isolation amplifier can power the sensor, transfer the signal from it, and isolate the two parts, preventing any currents from flowing across the barrier.

Removing the lid reveals some truly remarkable internal features. The centerpiece is a toroidal transformer. Unlike ordinary wirewound transformers, the upper half of each winding loop is an IC bond wire, while the lower half is a thick-film trace on the ceramic substrate.

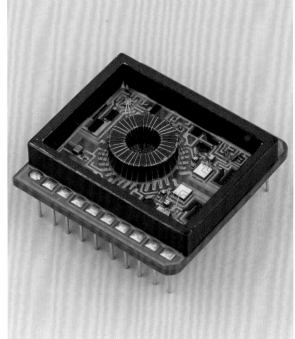

There are no electrical connections between the left and right halves of this amplifier module. The left side is the input side, and the right is the output.

Inside the Isolation Amplifier

Besides the transformer, this hybrid circuit features a few IC chips, diodes, and transistors. There are two ceramic chip capacitors, and quite a few laser-trimmed thick-film resistors.

A subtle feature of this device is that all of the internal surfaces—bond wires and all—are coated with a remarkably thin and uniform layer of clear insulating PARYLENE plastic. You can see it if you look closely at the bond wires and corners of the cubic dies. This is a vapor-deposited CONFORMAL COATING, which is grown onto all of the surfaces much in the same way that ice can form a perfectly uniform coating around branches and leaves during an ice storm. The coating provides general robustness and prevents high voltages from arcing between components and connections.

The square structure in the foreground
is a rather large individual transistor.
Other diode, transistor, and IC dies can
be seen in this view as well.

Afterword:
Creating Cross Sections

Many electronic components were harmed in the making of this book. In this section, we'll take you through each step in the process of creating the book's images, from cutting, cleaning, and mounting the components to shooting and processing the photographs.

Cutting and Polishing

Sawing, filing, milling, sanding: these are a few of the processes we used to open and prepare the various subjects. We approached each sample uniquely, examining it, cutting it apart experimentally, and then planning a final cut for maximum visual impact.

Our equipment included a low-speed diamond saw, diamond polishing discs, razor blades, power sanders, and a 5,000-kilogram milling machine, but the honest bulk of the work was carried out manually, with sandpaper, elbow grease, and patience.

We used small sheets of fine but conventional sandpaper, wetted with alcohol, on a highly flat surface. Depending on the nature of the subject, the final grit ranged from 600 to 10,000.

A diamond-coated disc cuts through the ceramic core of a carbon film resistor.

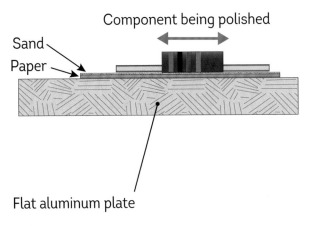

A solid carbide cutter in a milling machine makes a clean cut through the 3.5 mm audio jack. The jack is glued to a metal block to hold it firmly in place.

A low-speed diamond saw is typically used for materials analysis. Here, it makes an exploratory cut into an EPROM.

Cleaning

In addition to cutting, a substantial amount of prep time went into cleaning objects prior to photography.

Some vintage components came with well-preserved dust, trapped in coats of lacquer applied half a century ago. A more common issue was plastic, metal, ceramic, or semiconductor dust resulting from the cutting process itself.

For larger parts, we used compressed air and a clean, dry toothbrush, followed by a dust-cleaning gel. Tiny and sensitive parts were cleaned with a light spray of pure isopropyl alcohol, followed by drying with a compressed-gas duster. In one difficult case, we cleaned bond wires under a microscope with a brush consisting of a single cat whisker mounted to a handle of stainless steel tubing.

Even with these sometimes extreme measures, it remains the case that dust invisible to the eye can be seen at high magnification.

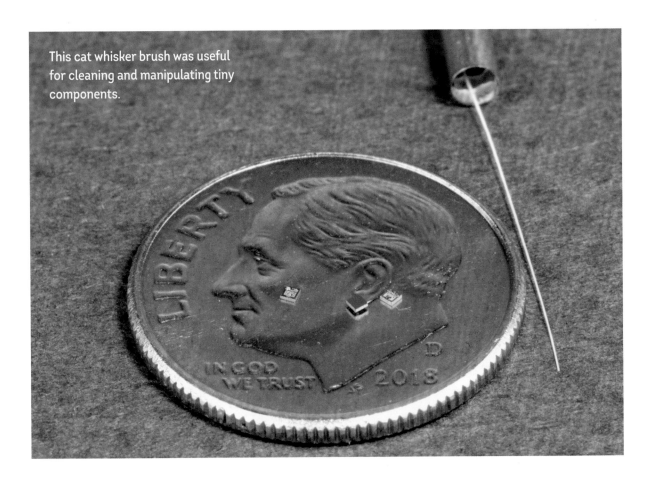

This cat whisker brush was useful for cleaning and manipulating tiny components.

Potting

Some complex or fragile samples would have fallen apart had we cut them on their own. A good example is the speaker (page 126), whose paper cone and thin voice coil wouldn't have survived cutting without something to hold them in place.

In these cases, we potted the subject in casting resin—a clear epoxy—to stabilize it during the cutting process. We used a small vacuum chamber to degas the mixed epoxy resin prior to casting, greatly reducing the number and size of air bubbles.

We tried to minimize the number of potted samples, since the results tended to be less clear, both literally and figuratively, than leaving empty space empty.

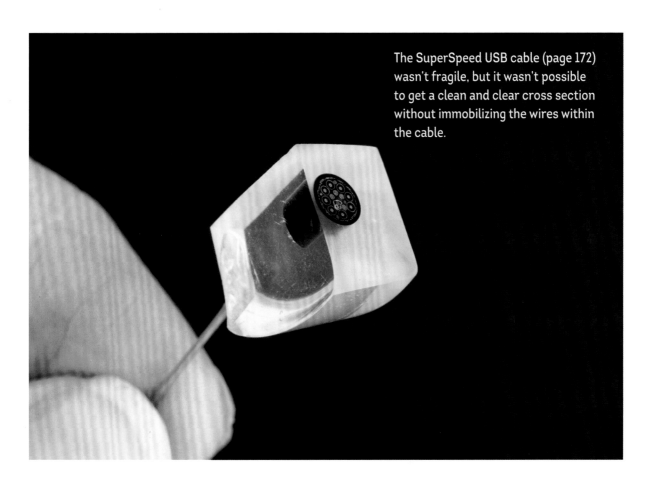

The SuperSpeed USB cable (page 172) wasn't fragile, but it wasn't possible to get a clean and clear cross section without immobilizing the wires within the cable.

Mounting

In many of our photos, the subjects seem to float in midair. This isn't the result of Photoshop or trickery, just careful positioning of the camera and subject.

We used a QuadHands-brand "flexible vise" with four alligator clips on gooseneck arms to hold many of our subjects in place, often with the clips just out of frame. For smaller objects or ones where the entire subject needed to be visible,

we glued them to metal support shafts. With carefully chosen angles, the shafts become nearly invisible.

The photo below shows the setup for our F-connector shot (page 155). Near the top left, an alligator clip grips the cable directly, while the connector is glued to a rigid stainless steel wire that passes through a piece of background paper. Behind the paper, two more alligator clips hold the wire in place.

These components are mounted for photography. Not all of them made the final cut.

Photographic Equipment

The images in this book are the result of conventional photography, with the primary exception that we used focus-stacking software.

We used the following camera equipment:

- Canon EOS 7D and EOS R camera bodies with lenses:
  - RF 24-105mm f/4-7.1
  - EF 100 mm f/2.8L Macro IS USM
- EF 28–135 f/3.5–5.6 IS USM
- MP-E 65mm f/2.8 1–5× Macro
- TS-E 24 mm f/3.5L II
- Two softbox flashes and a supplementary power flash with remote triggers
- Slik Pro 500DX tripod
- Custom linear focusing rail and shutter release based on AxiDraw hardware
- Software: Helicon Remote, Processing

A sixth lens–not listed–was used to take this particular photo.

The hardware beneath the camera is the custom linear focusing rail, along with cables for the camera power, shutter, and data readout.

Retouching

The photos in this book were collected and processed using Adobe Lightroom. Some consistent processes—what would have been called "developing" in the old days of darkrooms and chemicals—were applied to essentially every photo. These included corrections for the lens profile, as well as cropping, rotation, and adjustments for white balance, brightness, contrast, and overall tone.

Most photos had some amount of digital spot removal done within Lightroom. We used this tool to remove individual motes of dust and camera artifacts such as sensor dust, and to mitigate artifacts from sample preparation and general blemishes. An example is shown in the before/after photos below.

Heavy-handed retouching techniques such as digital "airbrushing" were used sparingly. We worked hard to preserve the true visual character of the objects presented here.

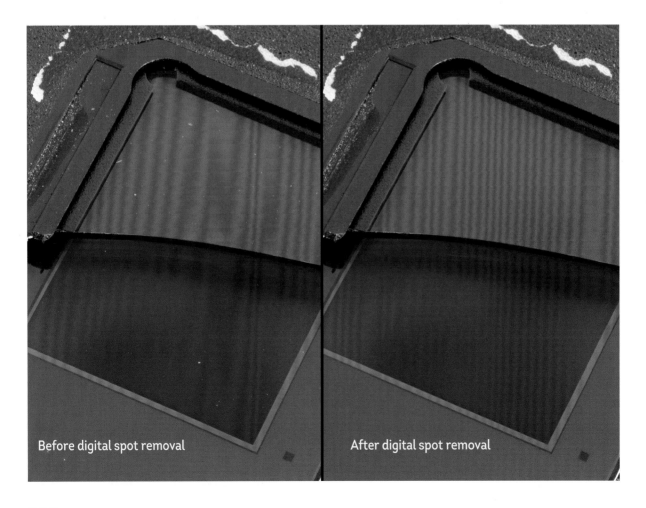

Before digital spot removal

After digital spot removal

About Macro Photography

The close-up images in this book are examples of MACRO PHOTOGRAPHY, defined loosely as photography where the presented image is larger than the actual subject.

In macro photography, the subject sits very close to the camera lens, and the range of distances that can be in focus tends to be very small. This situation is referred to as having a shallow DEPTH OF FIELD.

At our highest camera magnification, even with the camera aperture minimized to give the maximum possible depth of field, only about ¼ mm (0.01 inches) can be in focus at a time.

A shallow depth of field is such a well-known characteristic of macro photography that large-scale photos with a shallow depth of field may appear like miniatures to us—the so-called DIORAMA ILLUSION.

The diorama illusion makes this image, with a shallow depth of field, look a bit like scenery from a model railroad.

Focus Stacking

Many of the images in this book were processed with Helicon Focus. This software application specializes in **FOCUS STACKING**, a computational image processing technique that combines multiple images with a shallow depth of field to produce a single image with a greater depth of field. Focus stacking works by analyzing the images to identify the areas that are in focus, then stitching those areas together, similar to how software can stitch together photos to make a panorama.

Focus stacking can produce images with exceptional sharpness and depth of field, but achieving the best results requires a complex setup. Multiple images must be taken at equally spaced intervals and with equal exposures. When the depth of field is well under 1 mm, repositioning the camera requires great care. We used a robotic linear motion stage, along with custom software, to move the camera in precise, tiny increments and to take pictures at these locations.

This image is the result of combining eight equally spaced photos using focus stacking software. Slices through four of the eight originals are shown on the facing page.

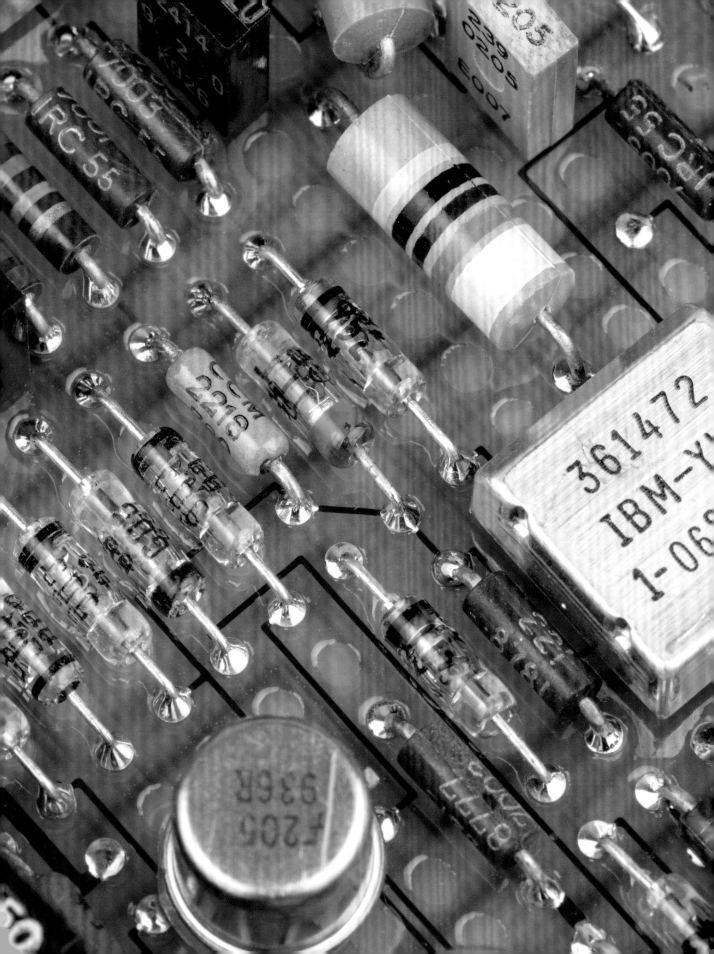

Glossary

ALTERNATING CURRENT (AC)

As opposed to direct current (DC), refers to electrical signals that smoothly cycle back and forth between positive and negative, pushing and pulling current as they do so. Most wall power outlets and power transmission systems use AC.

ANODE

The positive ("+") portion of a component that has both positive and negative parts. The negative portion is referred to as the *cathode*.

ANODIZATION

An electrochemical process, used with metals like aluminum and tantalum, that converts the outer surface into a textured metal oxide with a high surface area.

BAKELITE

A phenolic-based plastic resin frequently used as an insulator or housing in vintage electronics.

BALL GRID ARRAY (BGA)

A component package that makes connections to a circuit board through an array of tiny balls of solder on its underside.

BIMETALLIC STRIP

A thin metal structure made from two dissimilar metals with different rates of thermal expansion, such that it bends when the temperature changes.

BIPOLAR JUNCTION TRANSISTOR (BJT)

One of the basic types of transistor, typically used as an amplifier or switch for small electrical signals.

BOBBIN

A spool that wire is wound around.

BOND WIRES

Ultrathin metal wires that can attach directly to semiconductor dies.

BRUSHES

Spring-loaded assemblies in a DC motor that electrically connect the moving and stationary parts.

BRUSHLESS

Describes electric motors without brushes, which use rotating magnets instead of rotating coils.

CAPACITOR

A component that stores energy in the form of static electricity. Typically made of interleaved metal plates separated by a dielectric material.

CATHODE

The negative ("-") portion of a component that has both positive and negative parts. The positive portion is referred to as the *anode*.

CERMET

A composite material made from ceramic and metal.

CHIP
A colloquial expression for an integrated circuit, after "silicon chip."

COAXIAL
Describes items that share the same center axis.

COMPOSITE
A material or component made up of other materials or components.

CURRENT
The rate of flow of electrical charge through a circuit. Analogous to the rate of water flow (e.g., liters per minute) in a pipe.

DIE
The thin rectangular block of semiconductor material, typically silicon, that contains the active area of a semiconductor device. Multiple dies are sometimes referred to as *dice*.

DIELECTRIC
An electrical insulator. Different dielectrics may be chosen for their ability to withstand higher voltage or to increase the amount of electric field that can be stored, as in a capacitor.

DIODE
An electronic component that allows current to flow through it in one direction only.

DIRECT CURRENT (DC)
As opposed to alternating current (AC), refers to electrical signals where the current flows in one direction only.

Batteries and most plug-in power supplies output DC.

ELECTRODE
An electrical conductor that contacts a non-metallic part of a component such as a dielectric. It can also refer to electrical conductors that emit or collect electrons in a vacuum.

ELECTROLYTE
An electrically conductive fluid.

FERRITE
A ceramic filled with iron oxide.

FILTER
A device, analogous to an air or water filter, that allows only certain types of electrical signals or wavelengths of light to pass through it.

GERMANIUM
A chemical element; a semiconducting material similar to silicon.

HYBRID CIRCUIT
A component that consists of multiple other components, typically integrated circuits and passive components, along with a ceramic or fiberglass circuit board that connects them.

INDUCTOR
A component that stores energy in the form of a magnetic field. Typically made of copper wire wound around a ferrite form.

INSULATOR
A material that doesn't conduct electricity.

INTEGRATED CIRCUIT (IC)

A circuit made with many devices like transistors and resistors, fabricated together on a single semiconductor die.

INTERDIGITATED

Interleaved, as with interlocked fingers.

ISOLATION, ELECTRICAL

The transmission of power or signals between wires without an electrically conductive path connecting them. Sometimes referred to as *galvanic isolation*.

LASER DIODE

A special type of light-emitting diode (LED) that emits laser light.

LEAD FRAME

For integrated circuits packaged with pins, for example in DIP or SOIC packages, the set of metal forms comprising the external pins and connected to the IC through bond wires.

LIGHT-EMITTING DIODE (LED)

A diode designed to give off light when electricity flows through it.

MAGNET WIRE

Solid wire, typically copper, insulated with a very thin layer of varnish. Widely used in inductors, speakers, and other electromagnetic devices.

OSCILLATOR

A circuit element that produces an output signal at regular intervals or with a consistent frequency. Often used as the source for clock signals.

PHENOLIC

A class of chemical compounds, such as Bakelite, commonly used for the packaging of early electronic components and in circuit boards. Phenolic can also refer to certain composite materials, including circuit boards, even when they don't contain any phenolic-based resins.

PHOSPHOR

Various chemical compounds that emit visible light—some when struck by electrons, and some when struck by visible or invisible light. An important part of white LEDs, as well as vacuum tubes that light up, such as cathode ray tubes or vacuum fluorescent displays.

PHOTODIODE

A type of diode that produces an electrical signal when struck by light. Frequently used as a light detector, but also the basis of common solar panels.

PHOTOTRANSISTOR

A type of transistor that produces an electrical signal when struck by light and amplifies the resulting signal.

PLATED THROUGH HOLE

Holes drilled through a printed circuit board, where the inside surface of the hole is plated with copper to provide connections between the sides of the board. Vias are examples.

POLE

A contact terminal within a switch or relay.

POLE, MAGNETIC
Regions near the ends of a magnet where the magnetic field is the strongest, or pieces of ferromagnetic material in contact with those regions.

POTENTIOMETER
A three-terminal adjustable resistor, called a *pot* for short.

RECTIFIER
A diode used to convert AC to DC.

REDISTRIBUTION LAYER (RDL)
A miniaturized circuit board used as an alternative to a leadframe in integrated circuit packages that have large numbers of connections.

RESISTOR
A component that restricts the flow of electrical current and dissipates energy in the form of heat. Analogous to a narrow section of a water pipe.

ROTOR
The part of a motor that turns.

SENSOR
An electronic component that measures a physical property, such as temperature or light level, or that records an image, as in a camera sensor.

SHIELD
A barrier, usually of thin metal, used to reduce electromagnetic interference. It both reduces incoming signals and helps reduce emission of signals from within the shield.

SILICON
A chemical element; the semiconducting material most frequently used for fabricating integrated circuits.

SILICONE
A synthetic polymer made with a chemistry that includes silicon. Silicone rubber is a soft, rubbery material used for sealing components but also as bathroom caulking.

SOLDER
Any of various low-melting-point metal alloys used to establish electrical connections between components on circuit boards and sometimes within components.

SOLDER MASK
An insulating resin, often brightly colored, applied to circuit boards to control where solder can flow. It's the layer that makes most circuit boards look green.

SOLENOID
A coil of wire used as an electromagnet, for example in a relay or speaker.

SPRING FINGER
A spring-loaded electrical contact, arranged to make a consistent connection.

SPUTTERING
A precision process, carried out in a vacuum chamber, for depositing material onto a surface.

STATOR
The stationary part of a motor.

SUBSTRATE
The underlying surface that something is built upon.

SURFACE MOUNT
A method of connecting components by soldering them directly to one side of a circuit board.

TERMINAL
A connection point on a component, from where it can be linked to other circuit elements.

THICK FILM
A circuit fabrication technology that uses silkscreened conductive and resistive films that are fired like pottery glazes onto a ceramic substrate.

THIN FILM
A circuit fabrication technology based on etched patterns in ultra-thin layers of sputtered conductive or resistive materials.

THROUGH HOLE
A method of soldering components to a circuit board where the components have wire leads that go through holes in the board.

TRACE
An individual wire that's part of a circuit board.

TRANSFORMER
A solenoid made with more than one winding. Transformers are used for electrical isolation or to step up or down voltage with a tradeoff in current capacity.

TRANSISTOR
A three-terminal semiconductor component that allows one electrical signal to control another.

UNIVERSAL SERIAL BUS (USB)
A set of computer industry standards for cables, connectors, and protocols for communication between computers and peripherals.

VIA
Plated through holes connecting the copper layers of a circuit board, providing a path for signals that need to go from one layer to another.

VOICE COIL
The coil of wires in a speaker that moves when current is applied to it. Also refers to linear and rotary motors that work by the same principle.

VOLTAGE
A measure of the electrical potential available to move electric charge through a circuit. Analogous to water pressure in a pipe.

WINDING
An individual wire that may be coiled multiple times around a solenoid.

WIPER
A contact point that can move, as in the center terminal of a potentiometer.

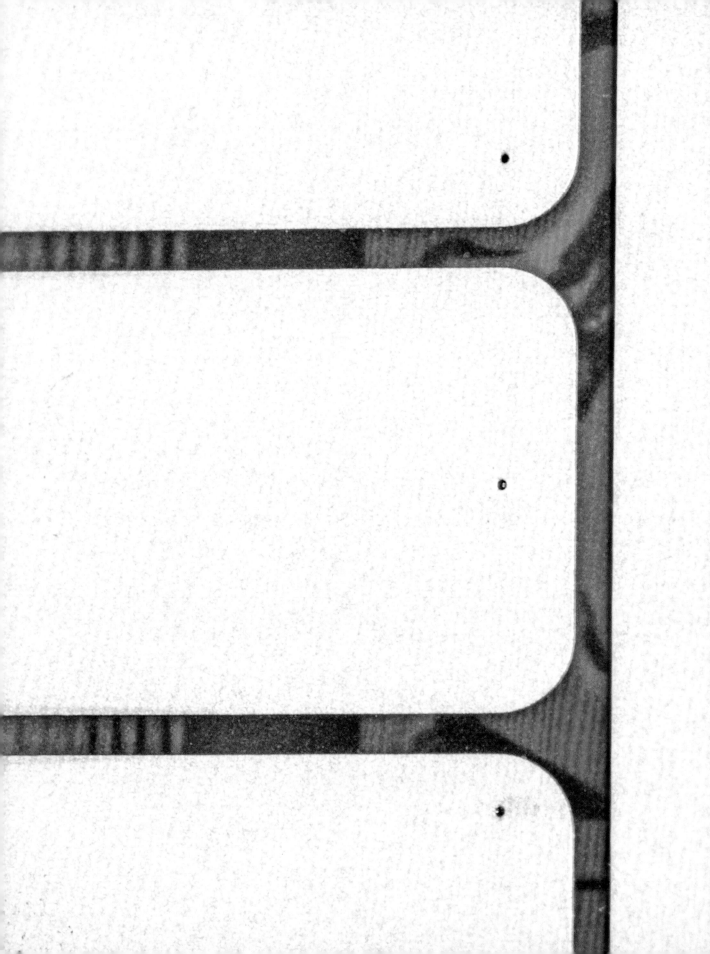

Index

Page numbers in **bold** refer to subjects in photographs.